How to Herd Clouds and Influence People

Published by Turbot Press, an imprint of Turbot HQ, Inc

The characters and events in this book are fictional composites based on real experiences. Names, organizations, and specific details have been changed to protect privacy and confidentiality.

First published online at cloudgovernance.org, July 2025
First print edition, 2025
Cover design by David Boeke

Library of Congress Control Number: 2025919710
ISBN 979-8-9931426-0-9 (paperback), 979-8-9931426-1-6 (ebook)

Published by Turbot Press
Turbot HQ, Inc
500 Westover Dr #20232
Sanford, NC 27330 USA

For digital resources and additional Cloud Governance materials,
please visit: www.cloudgovernance.org

For information about bulk purchases or speaking engagements: howtoherd@turbot.com

Praise for How to Herd Clouds and Influence People

"The Phoenix Project taught us to think in systems. How to Herd Clouds & Influence People teaches us to think in relationships. Both are essential for modern IT transformation."

— Jason Poley, Engineering Leader in Financial Services

"This book tackles the organizational challenges that are often missing from technical discussions about cloud governance. The narrative is delivered in an accessible way and resonates with the real enterprise dynamics I've seen. ... The path to technical solutions is always through people."

— Ken Robbins, President, Go2 Software; Former Executive Director of Engineering, Novartis Institutes for BioMedical Research (NIBR)

"In our rapidly evolving digital landscape, cloud governance has emerged as a critical strategic differentiator. [This book] offers leaders the strategic insights necessary to transform cloud governance from a compliance burden into a competitive advantage."

— Meredith Stein, co-author, Unlocking the Power of the Cloud

"Most tech books tell you what to do. This one shows you how to get people to actually do it. ... Just like The Phoenix Project became essential reading for anyone trying to understand and implement DevOps, How to Herd Clouds is essential reading for anyone trying to bring order to cloud chaos through effective Cloud Governance."

— Matty Stratton, DevRel Leader; Host of Arrested DevOps

"The author does a fantastic job of helping you get in the mindset of a change agent at a large organization, and some of the well-meaning pitfalls that change agent will fall into. It all comes down to aligning with people and finding common ground. At the end of the day, that other department isn't stupid, they just have their own goals, and the minute you find the commonality, you're off to the races! ... I highly recommend it for anyone wanting change in their organization."

— Michael Hedgpeth, Co-Founder, People Work; former Executive Director for Cloud Engineering at NCR; Observability Platform Leader, Splunk Enterprise Cloud

"One of the most unique and entertaining technical books I've ever read. The book does a brilliant job capturing an actual large enterprise cloud journey from manual processes and spreadsheets to self-service automated guardrails. ... This captures the real-world complexity of enterprise cloud governance that technical teams rarely see documented."

— Dave Shrestha, Enterprise Cloud Architect

"This story resembles many of my own efforts in the early days of almost every company I worked for; trying to fix the 'unfixable' and make a dent in the company's history! The details are spot-on and capture the real challenges of enterprise cloud transformation."

— Otto Aulicino, Sr. Director, Information Security, Laivly

"I found the examples from the book strikingly similar to the real world. Many companies don't realize that governance (including security) is missing until it was too late or they failed an audit or had a breach. Bob's book hits the nail on the head. I recommend this book for anyone who is trying to get their cloud landscape under control."

— Ravi D, Cloud Security Engineer

How to Herd Clouds and Influence People

Bob Tordella & Nathan Wallace

Turbot Press

Acknowledgments

To the visionaries who led enterprise cloud adoption,
and the skeptics who tried to stop it;
thank you for making this story possible.

Day 1: Access Denied

"Your account isn't set up yet."

Gary stared at the login screen, then at the sticky note with his temporary credentials, then back at the screen. Somewhere in Goliath Federal's maze of cloud accounts lay the infrastructure he was hired to manage. If only he could get in.

At his last company, this would have taken five minutes. A quick Slack message, an approval, done. But that was a different world - a 300-person tech company where everyone knew everyone meant Gary could just get things done when he needed to.

"Classic Goliath," chuckled Dennis, a developer who'd been showing Gary to his desk. "Took me three weeks to get access to our development accounts. By then I had missed contributing to my first assigned sprint. And that was just for Commercial Banking's cloud accounts," Dennis added. "Wait till you try accessing Retail or Wealth Management's environments."

"Three weeks?" Gary had known Goliath Federal would be different. That's why he'd taken the job as a Cloud Architect in the Enterprise Architecture team. After four years of running cloud operations at a startup, he wanted a bigger challenge. Looking at the error message, he was starting to understand just how big it was.

"You'll need to access the IT Access Management system to fill out Form 27-B," said a voice behind them. "Unless it's for Commercial Banking. That's 27-B-2. Or was it 27-C?"

Gary turned to find a woman with a "Zero Trust" sticker on her laptop studying him with mild amusement.

"Sarah," she offered. "Security. You must be the new cloud guy here to wrangle our three hundred-plus cloud accounts into submission?"

"Gary. And yes, though at the moment I'm just trying to..."

"Get access. I know." Sarah leaned against the cubicle wall. "Welcome to enterprise IT. We have ten different access processes, multiple approval steps, and hundreds of systems to complete a request. Nobody quite knows who owns what, but everyone's sure it's someone else's problem. Different teams, different tools, different priorities," she continued. "Each business unit does things their own way."

"Sounds like an opportunity," Gary said.

Sarah raised an eyebrow. "Sounds like chaos."

"Same thing, sometimes."

She almost smiled. "We'll see. Form 27-B is in the ITAM portal. Once you get access to that."

As Sarah walked away, Gary looked again at his blank screen.

At his last job, he could see everything, change anything, fix problems immediately. Here, he couldn't even log in.

Day 8: Found the Cloud Spend. Sort Of.

After a week of constant nagging, Gary was able to get access to one of the billing systems.

Gary opened the cloud billing dashboard for Commercial Banking's accounts. The numbers made him blink. Twice.

"Not what you expected?"

A woman at the next desk had noticed his reaction. Her laptop displayed a complex spreadsheet with highlighted cells and multiple tabs.

"I'm Fran," she said. "Finance. Been trying to reconcile those numbers for weeks."

"They seem..." Gary searched for a diplomatic word.

"High? Scattered? Impossible to track?" Fran supplied. "Yes to all three. And that's just Commercial. Wait till you see what the other divisions are spending."

She turned her laptop to show him. "Each division runs their own cloud expenses. Different accounts, different tracking tools, different approval processes. Some teams use reserved instances, others don't. Some tag resources, others..." she shrugged.

"No standardization?" Gary asked.

"We had standards. Multiple sets of them. Every division created their own." Fran's spreadsheet had a tab labeled "Standards Comparison v7.3."

Gary pointed to a particularly large number. "What's driving this spike?"

"The project team says it's for their new analytics project. But the tags don't match, the cost center is wrong..." Fran clicked through several spreadsheet tabs in rapid succession. "And of course, everyone's got their own naming conventions."

Gary nodded. Years of mergers and reorganizations had left their mark on Goliath's cloud infrastructure, just like its org chart.

"I've got finance teams in each division sending me different reports," Fran continued. "Each using their own tools, their own processes. Makes monthly reconciliation... interesting."

Day 15: Shadow Accounts

Gary was still digesting the cloud billing dashboard numbers from last week. After getting access to the primary billing system, he'd been working with Fran from Finance to understand Goliath's cloud spending patterns.

"I think I've got a handle on these 300 or so AWS accounts we're tracking," Gary said, tapping his pen on a notepad filled with calculations. "Commercial Banking's spending seems particularly high, but at least we can see where most of it's going."

Fran looked up from her computer with a knowing expression. "That's just the tip of the iceberg. I've been meaning to show you something else." She pulled a folder from her desk drawer and laid out several printed reports. "These are departmental credit card reconciliations from the last quarter."

Gary scanned the pages. Amid the normal business expenses were dozens of line items for cloud services, AWS charges, Azure subscriptions, even some GCP payments. Some teams were clearly running parallel cloud environments completely outside the corporate structure.

"Wait... all of these are additional cloud accounts?" Gary asked, flipping through the pages.

"Exactly. By my estimate, there are at least another 150 accounts we're not tracking formally. Some are small development instances, but others..." She pointed to a particularly large monthly charge. "This one from Commercial Banking's FinTech team runs about $20,000 a month."

Gary remembered his conversation with Dennis during his first week. "The developer who showed me to my desk mentioned missing sprint deadlines waiting for access. I'm guessing this is how teams are working around that?"

"Precisely," Fran confirmed. "When the official process takes weeks, but a credit card gets you running in minutes..."

Gary drummed his fingers on the desk. "But this creates huge blind spots, for security, for compliance, for cost management... for everything, really."

"That's why I wanted you to see this," Fran said. "The official AWS accounts we're tracking in the billing dashboard are easier to see, but these shadow accounts are where a lot of the actual innovation is happening."

Gary pulled out his phone and sent a quick message. Ten minutes later, he was sitting in Sarah's office in the Security Department.

"Shadow cloud accounts?" Sarah didn't seem surprised when Gary explained what Fran had shown him. "We've suspected as much. Some teams even told us directly they were using external accounts, but without a centralized inventory..."

She pulled up a security dashboard. "We're scanning and monitoring all the official accounts, but these credit card-funded environments are completely outside our visibility. We have no idea what controls they have, or don't have."

"Has anyone tried to bring them into the fold?" Gary asked.

Sarah shook her head. "Every time we approach a team about their shadow IT, they give the same response: 'If we move to the official accounts, we'll lose agility.' And honestly, they're not wrong. Our provisioning process is painfully slow."

Later in the day, Gary called Dennis to ask about the FinTech team's cloud usage. Dennis was surprisingly candid.

"Yeah, we've been running our own AWS environment for almost two years," he admitted. "Started because we needed to meet a deadline, but now our entire mobile banking platform runs there."

"Doesn't that create complications?" Gary asked.

"Sure, but it solves our biggest problem, getting things done quickly. The business wants features deployed weekly. The official cloud onboarding process takes a month." Dennis paused. "Look, we know it's not ideal. But we're delivering, and that's what matters to our leadership."

After the call, Gary stared at the credit card reports again. This wasn't just a handful of rogue developers. This was a systemic problem, one that wouldn't be solved by simply telling teams to stop using their credit cards.

Day 16: Ticket Maze

Gary couldn't stop thinking about the shadow accounts. Teams were bypassing official channels not out of rebellion, but necessity. He needed to understand why.

"Mind if I shadow you for a bit?" Gary asked Dennis the next morning. "I'd like to see your provisioning process firsthand."

Dennis looked surprised but nodded. "Sure. Actually, perfect timing. We need to set up a new development environment for a new module."

They settled at Dennis's desk as he pulled up a browser. "So here's how it usually starts," Dennis explained, navigating to an internal portal. "The official way, I mean."

Gary watched as Dennis logged into what appeared to be the main IT service catalog. The interface looked dated, with dozens of request categories nested in drop-down menus.

"Let's see... Infrastructure Requests... Cloud Services... AWS... New Environment," Dennis narrated as he clicked through the maze of options. "And now the fun begins."

The form that appeared made Gary wince. Dozens of required fields, many requesting technical details a typical developer wouldn't know.

"Network subnet preferences? IP address allocation strategy?" Dennis pointed at the screen. "Cost center and charge code? I have to tell you, I find some of this a bit much. I'm a developer, not a network engineer or accountant. I just need a place to build and test code."

Gary noticed the estimated completion time at the bottom of the form: 10–15 business days.

"Two to three weeks for a development environment?" Gary asked.

"If we're lucky," Dennis replied. "And that's just the first ticket." He opened another browser tab and pulled up a different system. "After that's approved, we need to submit access requests in the identity management system."

Another form appeared, equally complex.

"Next comes security approval," Dennis continued, opening yet another system. "I can never remember what this tool is called, but it's completely separate from the others. And don't get me started on databases. If we need one, that's an entirely different ticket to the Database Administration team in their own system."

By the time Dennis finished his demonstration, they'd navigated through five different ticketing systems across four separate portals, each with its own login, approval workflow, and user interface.

"How does anyone keep track of all this?" Gary asked.

"Most don't. They give up and use their credit cards," Dennis said matter-of-factly. "Or they request way more than they need upfront, because going back through the process for more resources takes too long."

Oscar from Operations was passing by but stopped abruptly when he heard the word "provisioning."

"Are you showing him our provisioning workflow?" Oscar asked, suddenly animated. "We've actually made some great improvements recently. Cut the average time down by two whole days last quarter."

"It still takes over two weeks," Dennis pointed out.

"That's impressively fast for enterprise infrastructure," Oscar said defensively. "You should have seen it a year ago. And remember, we're handling security, compliance, network integration..."

"But did anyone ever look at the entire process end to end?" Gary asked.

Oscar's enthusiasm dimmed. "Each team optimized their part without considering the overall journey. Security made their process more thorough. Operations made our reviews more detailed. Everyone did what made sense for their team."

"And the result is a process no one can actually use," Gary observed.

"Exactly," Dennis said. "So we have two choices: wait weeks to start work or they charge it to their corporate account."

"What would make you use the official process?" Gary asked.

Dennis thought for a moment. "Obviously, it needs to be faster and simpler. But there are other things too. When we run our own accounts, we spend days requesting and then configuring VPN connections to internal systems, figuring out how to hook into an identity system, and we just rely on default networking configurations. And, painfully, every month I have to file expense reports and justify the charges."

Dennis sat forward. "If the official process gave us pre-configured access to internal networks, handled the identity management, and just took care of billing automatically, all while being reasonably fast, that would be worth switching for. We want to build applications, not manage service requests."

Later that day, Gary mapped out the entire cloud provisioning journey on a whiteboard in his office. The resulting diagram looked like a complex subway map, with lines crossing, diverging, and occasionally leading to dead ends.

Day 22: Inventory Quest

A week after exploring the ticket maze with Dennis, Gary was still troubled by a fundamental question: what resources did Goliath actually have in the cloud? His growing collection of cloud cost reports raised more questions than answers.

"You must be Gary." A man with ServiceNow stickers on his laptop walked up to his desk. "Adam from Asset Management. Heard you've been asking about our cloud inventory."

"Fran mentioned you might help me match resources to costs," Gary said. "The numbers just don't add up."

Gary turned his screen to show a spreadsheet. "According to billing, Commercial has over a thousand EC2 instances. But I can only see about four hundred in the console."

"Ah." Adam nodded knowingly. "Welcome to my world. Check this out." He opened his laptop to show a dashboard. "This is our Configuration Management Database, our CMDB. According to this, Commercial has exactly six hundred and twelve servers."

"But the billing shows..."

"A thousand? Yeah. And tomorrow it might show two thousand. Or five hundred." Adam scratched his head. "Resources come and go from auto-scaling, development testing, and batch processing. Resources appear and vanish faster than we can track them."

Gary thought back to his old company. "But the cloud itself is already a virtual representation of what's running. Why are we trying to create another copy of that data instead of just querying it directly?"

"In theory, that's what we could do," Adam replied with a hint of defensiveness. "But you have to understand, we're dealing with inventory processes originally built for physical servers. The serverless platform services are edge cases. From my understanding, most enterprises don't even attempt to track those yet."

"How do teams keep track of all the resources they're running?"

"Depends who you ask. Development has their infrastructure as code. Operations have their monitoring tools. Security has their scanners. Each business unit has their own..."

"Let me guess," Gary interrupted. "Their own tools?"

Back at his desk, frustrated with how fragmented everything was, Gary was determined to understand the full picture.

Gary arranged quick meetings with other teams that same day. Sarah from Security showed him their vulnerability scanning dashboard.

"We scan everything we know about," she explained, "but that's the problem, we don't know what we don't know. When we discover a new critical vulnerability, I can't tell you with confidence if we've patched everything."

In the break room, Oscar shared his frustrations.

"We have monitoring for the cloud resources we know about, but teams are constantly deploying new services without telling us. Last month, we had an outage in a critical service running in a shadow account. It took us hours to even find it."

Oscar shook his head in frustration. "And frankly, we can't even trust what's in the CMDB anymore. The data is so inconsistent that my team has started keeping their own inventory of production systems. It's the only way we can be sure what we're responsible for."

By the end of the day, Gary's head was spinning. He had three completely different counts of cloud servers from three different sources. Multiple tools tracking the same environment. Different teams are maintaining separate inventories because they couldn't trust the official one. A chaotic picture where nobody had a complete view.

Gary mapped the situation in his notebook, drawing lines between teams, tools, and data sources. The result looked like a tangled web with no center.

Day 30: Sarah's View

Sarah took a sip of her coffee as she scrolled through the latest batch of security alerts. Another day, another hundred notifications about misconfigured cloud resources. She'd been the security lead for Goliath's cloud environments for two years, and some days it felt like she was the only one who could see the disaster waiting to happen.

Her phone buzzed with a calendar reminder: "Cloud Security Sync with Gary."

It had been a month since Gary had arrived with his ambitious ideas about bringing order to Goliath's chaotic cloud landscape. She liked him, he was sharp and didn't give up easily. But she'd seen this movie before, and it rarely had a happy ending. Usually the ambitious newcomer either quit in frustration or eventually gave up and accepted the status quo.

Sarah gathered her laptop and headed to the small conference room. She arrived early and pulled up the dashboard she'd prepared. The numbers weren't pretty. Over 70% of Goliath's cloud accounts had critical security issues. Especially with identity and access management configurations. IAM roles with excessive permissions had increased by 28% in the last quarter alone. And that was just in the environments they knew about.

Gary arrived, looking more tired than when he'd started.

"Thanks for making time," he said, settling into a chair. "I wanted to understand how Security sees our cloud challenges."

"Honestly? As a ticking time bomb," Sarah replied, turning her laptop around. "This is our current cloud security posture."

Gary studied the dashboard more closely. "That's... concerning."

"That's diplomatic of you." Sarah pointed to a particularly alarming graph on her CNAPP security platform dashboard. "This shows the proliferation of IAM roles. Two years ago, we had about 200 roles across all our accounts. Today, we're over 3,000, most with permissions far beyond what they need. Classic permission inflation."

"Why haven't these been restricted?" Gary asked.

Sarah laughed. "You've met Dennis, right? His team is just one of dozens who push back whenever we try to limit permissions. "We can't release it on time if we have to wait for Security approvals. And honestly, they're not wrong. Our processes are too slow."

She closed her laptop. "Look, I've been here five years. I've watched three different attempts to centralize cloud controls. All failed for the same reason: they tried to control instead of enable."

"What do you mean?"

"Central IT has a reputation here. When teams hear about centralization efforts, they translate it to 'slow me down.' And they've got the receipts to prove it." Sarah counted off on her fingers. "Two years ago, the Enterprise Architecture team tried to standardize cloud deployments. Their approval process took so long that teams just ignored it."

She continued, "Now Operations has shifted to being 'cloud brokers,' where you request an account through them. But as you've seen, they still can't keep up with requests, and teams are working around the process by creating their own accounts. And let's not forget when my own team tried to implement security reviews for all cloud resources."

"What happened?"

"Teams submitted the bare minimum documentation, we approved based on incomplete information, and everyone pretended the process added value." Sarah shook her head. "That's why teams don't trust central IT initiatives anymore. We've cried wolf too many times, promising to make things better but just adding bureaucracy instead."

"So what would work?" Gary asked.

Sarah thought for a moment. "If you want my honest opinion? Stop trying to control the cloud. Start enabling teams to use it better. Dennis doesn't want to bypass Security; he wants to deliver features without unnecessary delays. The business units don't want insecure environments; they want their teams to maintain their velocity."

As Gary gathered his things to leave, Sarah's tone softened. "Listen, I don't mean to sound so cynical. I actually hope you succeed where others couldn't. We desperately need someone to break this cycle." She paused. "Just... be careful about rolling out another top-down IT initiative. That's exactly what everyone is afraid of."

After Gary left, Sarah turned back to her alerts queue. Another IAM role with admin permissions had just been created. Another AWS S3 storage bucket publicly exposed. Another network security group with wide open firewall rules to the internet. She scrolled through the seemingly endless list, feeling the familiar weight of futility.

Their CNAPP tool was excellent at finding security problems, but completely useless at getting them fixed. Every day, the backlog grew larger. Every day, the teams created more issues than they resolved. At this rate, they'd need to triple the Security team just to keep track of all the findings, let alone remediate them.

She hoped Gary saw the bigger picture. They were drowning in technical issues, but those were just symptoms. The real challenge wasn't finding the problems; it was getting anyone to fix them. They might not have visibility into everything happening in their cloud, but what good was visibility when you couldn't drive change?

Day 38: Perfect Opportunity

Gary studied the strange patterns in the cloud billing report. After weeks of digging through cost data with Fran, they'd uncovered something promising.

"Look at this section," Gary said, highlighting a row in the spreadsheet. "AWS EBS storage volumes with no instance attachments. Hundreds of them, just sitting there, incurring charges."

Fran leaned in for a closer look. "You're right. They're not attached to anything, but we're still paying for them every month." She did a quick calculation. "These unattached volumes alone are costing us over $30,000 a month across the accounts we can see."

Gary did the math. "That's over $350,000 annually for storage nobody's using. And this is just what we can see in the accounts we have access to."

"Low-hanging fruit," Fran replied. "No impact on applications, no risk to production, just pure waste."

The numbers were compelling, but Gary had learned that data alone wouldn't drive change at Goliath. He needed a proof of concept, a success story to demonstrate what was possible.

"What I need is a team willing to test this," Gary said. "Someone open to trying a cleanup process."

"That might be easier than you think," Fran replied. "The quarterly business review is coming up in six weeks, and Finance is under pressure to find cost savings. If you can show concrete results by then, you'll have everyone's attention."

Gary thought for a moment. "Who might be receptive? I need someone technical enough to implement the changes, but also open to working with central IT."

"Peter's team from Payments," Fran suggested immediately. "They run the payments platform, so they have the clout to try something new. And Peter's always looking for ways to redirect resources from infrastructure to feature development."

Two hours later, Gary was sitting in Peter's office, walking through his proposal. Unlike many of the other technical leads, Peter didn't immediately bristle at suggestions from central IT.

"So you're saying we're paying for volumes that aren't even being used?" Peter asked, studying the spreadsheet Gary had prepared.

"Based on the billing data for your accounts, yes," Gary confirmed. "We've identified at least 50 unattached volumes in your development and test environments."

Peter nodded thoughtfully. "We've talked about implementing a cleanup process before, but it keeps getting pushed down the priority list. There are always other fires to put out."

"What if I helped build the automation?" Gary offered. "A simple script to identify and tag resources for review, with the option to clean up after approval."

"And what's in it for you?" Peter asked, not unkindly.

"Data," Gary replied honestly. "I need to prove this concept works before I can suggest it more broadly. Your team would be helping establish a pattern others might follow."

Peter considered this. "I like the idea of being first. And frankly, if this works, it frees up the budget that I could use for new features instead of paying for digital garbage." Peter nodded decisively. "Let's do it. My team can help identify what's safe to remove."

As Gary walked back to his desk, his mind was already racing with ideas for the automation. This was exactly the kind of opportunity he'd been looking for, a clear problem with a straightforward solution that could demonstrate immediate value.

Day 45: Early Adopters

"I think this should work," Gary said, "But I'm not a developer. I need someone to check my logic."

Gary reached out to Oscar. He wasn't a proper developer either, but he'd written enough automation scripts for Operations to spot potential issues. Oscar studied the Python script on Gary's screen. "The basic idea is sound. You're querying for unattached volumes, then tagging them for review. After your grace period, you'll snapshot the volumes, then delete anything still tagged." He frowned slightly. "But there are some edge cases you're not handling."

Gary sighed. "Like what?"

"What if a volume gets reattached during your grace period? You need to automatically remove the deletion tag. And what happens if someone removes your tag but doesn't actually reattach the volume? You'd miss it in the next cleanup cycle."

These were exactly the kinds of details Gary was worried about. Writing a script that worked in a test environment was one thing; creating production-grade automation that wouldn't accidentally delete important resources was another challenge entirely.

"I wish we had a developer he could work with," Gary admitted. "This would take them a day instead of a week. Even with me using AI to assist, we need more technical skillsets on the team for this kind of work."

After several more days of coding, testing, and refining, Gary finally had something he felt comfortable sharing with Peter's team. He'd solved the basic scenarios, but knew the script would need more refinement as they found edge cases.

"It's not perfect," Gary explained as he walked Peter through the code, "but it should safely identify resources for cleanup while giving your team the final say on what actually gets deleted."

"Let's try it," Peter said. "We can run it on our development account first with minimal risk there."

They set up the scheduled task to run the script weekly, with a three-day grace period before deletion. The first execution identified 320 unattached volumes.

Peter's team reviewed the list and confirmed that most were indeed abandoned resources. They found a few false positives, volumes temporarily detached during maintenance, and removed the deletion tags from those.

When the grace period expired, the script cleaned up the remaining resources, instantly saving the team over $5,000 per month in unnecessary charges.

"This is exactly what we needed," Peter told Gary at their review meeting. "Simple, effective, and it doesn't require us to change how we work. Any chance we could get this running on our other accounts?"

"Absolutely," Gary replied, feeling a surge of optimism. "We'd just need to adjust the parameters for each environment."

As word of the successful pilot spread, other teams began reaching out. Not the resisters or the skeptics, but other early adopters eager to eliminate waste in their environments.

Jennifer from the Retail Banking team was first. "Peter told me about your cleanup script. We've got at least a hundred unattached volumes in our sandbox accounts. When can we get this running?"

Then came Dinesh from the Data Analytics group. "Heard you helped Peter clean up unattached volumes. No doubt we have thousands of them that were never cleaned up over time. Can your script handle that volume?"

By the end of the week, Gary had requests from six different teams, all willing to try his automation in their environments.

But as he began deploying to these new accounts, complications arose. Each team had different tagging standards. Some needed different retention periods for regulatory reasons. One team needed emails sent to a distribution list instead of individual users.

What had started as a simple script was quickly evolving into something more complex. Gary found himself adding configuration options, exception handling, and team-specific customizations. Each new feature made the script more powerful but also more difficult to maintain.

"This is getting unwieldy," he admitted to Fran. "I'm spending all my time customizing and maintaining this script instead of finding the next opportunity."

"But it's working," Fran pointed out. "These teams are seeing real savings. Peter's monthly costs are down 12% already."

She was right. Despite the challenges, they were making progress. Teams were seeing tangible benefits from working with central IT rather than avoiding it.

Day 52: Barry's Kingdom

"The Commercial Banking Division isn't interested."

Gary stared at the email, reading it for the third time. After the success with Peter's team and several other early adopters, he'd sent a proposal to extend the cost savings initiative across all divisions. Most had responded positively or at least asked for more information, except Commercial Banking.

Their response was just six words, a blunt dismissal.

"Did you see this?" Gary asked Fran, showing her the email.

She grimaced. "Ah, you've met Barry. Well, not really met him, but felt his presence."

"Barry?"

"Barry Swinton, Commercial Banking's IT director. Twenty years at Goliath. Runs their IT function like it's his personal kingdom. It's always his way or the highway, and he makes sure he takes all the credit for any wins." Fran sighed. "His division generates almost a third of the bank's profits, so leadership gives him a lot of runway."

Gary scrolled back through the email. "But this makes no sense. We've proven this works. They could save hundreds of thousands with minimal effort."

"Around here, Barry's always the exception to every rule and every initiative," Fran replied. "He only participates if it was his idea first."

Gary wasn't ready to give up. "I need to meet with him. This is too much potential savings to ignore."

Two days and several emails later, Gary had a meeting on Barry's calendar. Getting fifteen minutes had taken more effort than scheduling time with the Head of Technology Operations.

Barry's office was on the executive floor of Commercial Banking's building, a ten-minute walk from the main IT department.

"Gary, is it? Come in." Barry gestured to a chair across from his imposing desk. In his mid-fifties, with a tailored suit and carefully combed silver hair, he exuded confidence and authority. "I've got another meeting in twelve minutes, so let's be brief."

"Thanks for making time," Gary began. "I wanted to discuss the cost savings opportunity with unattached volumes in your cloud environments."

Barry nodded. "I read your proposal. Very... enthusiastic."

"We've implemented this successfully with six teams already. They've seen immediate savings with zero operational impact."

"Good for them. But Commercial Banking manages its own cloud resources. Successfully, I might add. We were in AWS two years before central IT even had a cloud strategy."

"That's impressive," Gary acknowledged. "But that's exactly why there's so much opportunity. According to our analysis, you could..."

"Save approximately $240,000 annually by eliminating unattached volumes," Barry finished. "Yes, I've read the numbers."

"Then I'm confused about why you're not interested."

"Why we're not jumping to implement your solution?" Barry straightened in his chair. "Let me be clear. Commercial Banking IT has delivered fourteen consecutive quarters of on-time project delivery. We've maintained five-nines availability for our customer-facing systems. We've launched three major platforms ahead of schedule."

He gestured to a wall of awards and recognition. "We've done this by controlling our own destiny. Our systems, our rules. It works. My team knows our business, our compliance requirements, our customer needs."

"I'm not suggesting taking over your operations," Gary clarified. "Just implementing proven cost savings."

"With your script. In our environment. Managed by your team." Barry shook his head. "Each new dependency on central IT is a new point of failure. A new approval process. A new way for someone outside my control to impact our delivery."

He checked his watch. "I've got to prepare for my next meeting. My team will continue optimizing our cloud costs using our methods, on our timeline. I appreciate your initiative, but Commercial Banking will handle this internally."

As Gary walked back to his building, he felt a mix of frustration and confusion. The rejection wasn't based on technical merits or even concerns about the approach. It was about control, territory, and a deeply ingrained resistance to anything outside of Barry's full control.

Gary did some digging. Commercial Banking had been an early cloud adopter, using AWS to quickly launch new digital banking services while central IT was still focused on data centers. They'd built their own Cloud

Team, created their own processes, and established their own vendor relationships. Their success had become intertwined with their independence.

The numbers told an interesting story as well. Commercial Banking's cloud costs were 40% higher per workload than other divisions, but they consistently delivered projects faster than other divisions. They were paying a premium for speed and control. And from their perspective, it was worth every penny.

Sarah from Security dropped by. "So, you met the great Barry Swinton?"

"That obvious, huh?"

She smiled sympathetically. "We all go through this rite of passage. Last year, I tried to get his team to implement our cloud security standards. His exact words were 'we've been handling our own security since before your team existed.'"

"But his environments have the same issues as everyone else's," Gary pointed out.

"More, actually," Sarah said. "But they also have more resources and more autonomy to address them their way. From their perspective, working with central teams just adds bureaucracy."

Gary considered this. "There's got to be a way to add value without threatening their autonomy."

"If you figure that out," Sarah replied, "you'll have solved the puzzle that's stumped every central team for years."

Day 60: Dennis's Dilemma

Dennis closed the email about Gary's cost optimization efforts and stared at his screen. The proposal made perfect sense, a simple automation to clean up unattached volumes and reduce cloud waste. Peter's team was already using it with good results. His own team should probably do the same.

And yet, he couldn't bring himself to reply.

It wasn't that saving money was a bad idea. But adopting it now meant admitting something uncomfortable: they'd been wasting money all along. Money they could have saved months ago if they'd prioritized it.

He glanced at the whiteboard across from his desk, where he'd mapped out the next six sprints for the mobile banking app. User authentication improvements, expanded bill pay features, integration with the new rewards system, all with aggressive deadlines that his team was already struggling to meet.

Three years ago, when they launched the first version of that app, Dennis had made a career-defining decision. After weeks of waiting for central IT to provision their environments on-prem, he'd pulled out his corporate card and spun up AWS accounts and resources himself. What had been projected as a six-month delivery became six weeks.

His director Barry had praised his initiative, and the team became known for getting things done despite obstacles. Which had accelerated his team's use of their own cloud environments. Even Barry had nodded approvingly when Dennis explained how they'd circumvented the slow central processes to meet business deadlines.

But that independence had created its own burden. His phone buzzed with a message from one of his developers.

"Another auto-scaling failure overnight. I fixed it, but it took 3 hours I should have spent on the rewards integration. We really need dedicated SRE operational support."

Dennis winced. His developers were spending almost a third of their time managing cloud resources instead of building features. This was time they couldn't afford to lose, considering their current deadlines.

He pulled up their AWS bill, 15% over budget again. His team had identified cost optimization opportunities months ago, and even created issues in GitHub to track them. Ultimately, they kept getting pushed down the priority list.

He started to compose an email to Gary about the cost optimization initiative, then stopped mid-sentence. What would Barry think? The Commercial Banking IT Director had been adamant about his position: they handled their own infrastructure, period. Reaching out to central IT might be seen as undermining that stance.

But every hour spent managing infrastructure was an hour not spent delivering features. And with their current project backlog, they couldn't afford to waste either time or money.

Dennis deleted his draft email to Gary. It was easier to keep things as they were. Easier not to risk changing a model that had brought him recognition. Easier not to give central IT a "win" after years of working around them.

Dennis closed his laptop and stared at the whiteboard with their upcoming deadlines. Working with central IT meant giving up some control. But maintaining complete independence was becoming increasingly costly in ways that weren't just financial.

Maybe after this release, he thought, fully knowing that the next release would bring its own pressures and priorities.

Day 75: Wall of No

"The cost savings initiative isn't getting traction with the larger divisions."

Gary looked up from his laptop as Fran dropped into the chair across from his desk. After the successful pilots with Peter's team and a handful of others, they'd presented the results to the rest of the organization, expecting broader adoption. Instead, they'd hit a wall.

"What's the pushback?" Gary asked, though he had a good guess.

Fran placed a folder on his desk. "Here are the responses. Commercial Banking we already knew about; Barry shut that down immediately. But even Retail and Wealth Management are hesitant."

Gary flipped through the emails. The objections varied in wording but followed the same pattern: "We'll handle this ourselves." "Our environment has unique requirements." "We're already working on a similar solution internally."

"This doesn't make sense," Gary said. "We've proven this works. The data shows that Peter's team saved 12% in the first month alone. The script is free, it's minimally invasive, and it gives teams complete control over what gets cleaned up."

"It's not about the technical merits," Fran replied. "It's politics."

"Explain."

"Most of these divisions have been building their cloud capabilities for years, long before central IT had a coherent cloud strategy. They see any central initiative as a threat to their autonomy, regardless of how much sense it makes."

Gary thought about his conversation with Barry, his exploration of operations silos with Oscar, and Dennis's reluctance to engage. The pattern was becoming apparent. This wasn't about the script or the cost savings; it was about history, trust, and organizational dynamics.

"I have my check-in with Charles tomorrow," Gary said. "Since he brought me in specifically to tackle these cloud challenges, I need to bring this up."

Fran looked concerned. "Are you sure? Escalating to the Head of Technology Operations might be seen as going over their heads."

"Charles needs to know that we've identified significant savings but are hitting resistance. These are exactly the kinds of issues he hired me to address."

The next day, Gary sat in Charles's office for his scheduled review. After discussing his initial findings and the successes with Peter's team, Gary raised the issue of resistance from the larger divisions.

"Just to confirm," Charles said after reviewing the data, "you've identified nearly $1 million in annual savings through this cost optimization initiative?"

"That's a conservative estimate with just a few cost optimizations," Gary confirmed. "And we've already verified the approach with several teams who are seeing immediate results."

"And the primary resistance is coming from... where, exactly?"

"Some divisions prefer to handle cost optimization on their own. They have concerns about central IT initiatives potentially impacting their operations."

Charles sat back, visibly frustrated. "So they're rejecting or dragging their feet on a proven cost-saving measure, and they've been wasting money this whole time."

"That's an accurate assessment," Gary said.

"I see." Charles made a note. "This timing works out well. I've been planning to get the division IT leaders together for a business review next week. I'd like you to present your findings there."

As Gary walked back to his desk, he felt uneasy. He'd hoped for Charles's support, but turning this into a formal presentation to all division heads changed the dynamic. It was no longer about collaboration; it would be perceived as a challenge.

The next day, his phone rang. It was Barry, and he wasn't happy.

"I understand you raised concerns about our cloud spending with Charles," Barry said without preamble. "That was a mistake."

"I presented our findings during my scheduled review," Gary replied carefully. "The potential savings are significant across all teams."

"Commercial Banking manages its own cloud costs. We have our own optimization roadmap, our own priorities, and our own methods. Going to Charles hasn't changed that, it's just made me question whether you can be trusted as a partner."

After Barry hung up, Gary sat back in his chair, conflicted. The data supported their initiative. The approach was sound. The savings were real. But in pushing it through channels, they might have damaged the very relationships they needed to succeed long-term.

Later that day, Sarah popped in. "Heard you woke the dragon," she said with a sympathetic smile. "Barry's been on the warpath all morning."

"I discussed it with Charles during my review," Gary defended. "The savings are too significant to ignore."

"I'm not saying you're wrong," Sarah replied. "But in an organization like Goliath, being right isn't always enough. You need buy-in, not just a mandate."

Day 79: Divided We Stand

Gary reviewed his slides one last time before the divisional IT leaders' meeting. Charles had given him fifteen minutes to present the cost savings initiative and its results. Easy to understand data, compelling examples, straightforward next steps; it was all there.

"Ready?" Charles asked, poking his head into the conference room where Gary was setting up.

"As I'll ever be," Gary replied, trying the third dongle to finally get his laptop connected to the projector.

The room filled quickly with division IT directors and their deputies. Barry arrived last, taking a seat directly across from Gary, his expression carefully neutral.

Charles opened the meeting with a brief overview of cloud spending across the organization. "We're consistently over budget on cloud resources, which is why I've asked Gary to share his findings on a cost optimization initiative his team has been piloting."

Gary walked through the data methodically. The unattached volumes, the cost of those resources, the simple automation that had already saved Peter's team thousands. The presentation was factual, data-driven, and deliberately non-confrontational.

"In summary," Gary concluded, "we have a proven approach that's already working with several teams. We'd like to extend this to all divisions and could help implement it within the next two weeks."

The room was silent for a moment before Barry spoke. "This is certainly... interesting data. But I'm confused about something." He sat forward. "Is central IT now responsible for divisional cloud costs?"

"We're offering a solution..." Gary began.

"A solution implies we have a problem that needs solving," Barry interrupted. "Commercial Banking manages our costs according to our own priorities. Some resources you're labeling as 'waste' are kept for specific reasons, such as disaster recovery, testing, and compliance requirements."

"The data shows..." Gary tried again.

"The data shows that you've looked at our cloud resources without context," Barry said. "Without understanding why those resources exist in the first place."

Gary could feel control of the meeting slipping away. The Retail Banking director jumped in next, followed by Wealth Management. Soon, they were all talking over each other, each highlighting a different reason why the initiative wouldn't work for their division. What should have been a discussion about implementation had devolved into a chorus of objections.

Charles stepped in. "The savings are substantial enough to warrant consideration. What would you need to move forward?"

This shifted the conversation slightly. The division heads began listing conditions: more division-specific analysis, guarantees against disruption, detailed reviews by their own teams first, implementation only after their approval for each resource.

Gary's head was spinning. What was supposed to be a quick, simple cost savings win for the organization had somehow spun into a huge project indirectly related to the original idea. How had a straightforward automation script morphed into this complicated enterprise initiative?

"So we have agreement in principle," Charles summarized, "with implementation pending further review by each division." He glanced at Gary. "Gary's team will work with each of you to provide the additional context you've requested."

As the meeting dispersed, Barry lingered. "A word of advice," he said, his voice low. "Next time you want to analyze my division's resources, talk to my team first. Context matters."

Day 80: Second Thoughts

Gary replayed the meeting in his mind. What had gone wrong? The data was sound. The approach was proven. But he'd underestimated how protective divisions would be of their autonomy, and how easily they could delay action by demanding more analysis.

Fran showed up at his desk. "How did it go?"

"We got 'agreement in principle' and a list of barriers that will keep us from actually doing anything," Gary replied. "They all want more context, why resources exist, who owns them, and what business function they serve."

"That's the tagging problem," Fran said. "Without consistent resource tagging, you can't tell what anything is for or who owns it."

Gary thought about this. "So we need better tagging to provide context for cost optimization?"

"We need better tagging for everything," Fran replied. "But good luck with that. Every division has their own tagging standards, or none at all."

The next day, Gary began digging into the tagging practices across divisions. What he found was even worse than Fran had suggested. Some teams used detailed tagging schemas. Others tagged sporadically. Many didn't tag at all.

Without consistent tagging, providing the "context" divisions demanded was nearly impossible. How could you determine which unattached volumes were truly unused versus temporarily detached for maintenance? Which development instances could be safely shut down after hours versus those running long-term tests?

Gary arranged a call with Peter to see how his team handled this.

"Oh, we tag everything," Peter explained. "Owner, environment, application, and cost center. It's all automated through our deployment pipeline. Makes it easy to identify resources and their purpose."

"But that's not standard across Goliath?"

"Not even close. Every team does it differently. Most likely impossible to get a consistent view across divisions."

As Gary dug deeper, he realized the tagging inconsistencies weren't just complicating his cost initiative. They were a symptom of a deeper problem. Without a common language to describe cloud resources, every analysis, every optimization, every security assessment would face the same barrier.

Day 82: Tag Wars

"temp-do-not-delete. Is this really a tag in production?"

Gary stared at his screen in disbelief. He'd been reviewing resource tags across Goliath's cloud accounts for three days, building a case for standardization. What he'd found was digital chaos.

"That server's been running for six months," Fran said, looking over his shoulder. "Monthly cost of about $500."

"And no one knows what it's for?" Gary asked.

"Well, someone naming it 'temp-do-not-delete' implies they knew at some point." Fran pulled up another screen. "But look at this monthly cloud cost review. See how many different values we have just for the 'environment' tag?"

Gary scanned the list: "prod, production, PROD, prd, Prod-env, live..." The variations seemed endless.

"Without consistent tagging, we're flying blind," Fran continued. "Finance can't allocate costs properly. Security can't assess risk accurately. Operations can't prioritize support correctly."

Gary had scheduled a meeting to discuss tagging standards with cloud engineers and architects from each division. After the divisional leadership meeting debacle, he'd decided to start with the practitioners who actually implemented the tags, people who understood the technical challenges without the political baggage.

He'd asked Peter and Dennis to recommend technically-minded colleagues, resulting in a focused group of about eight cloud engineers and architects.

The conference room filled quickly. This topic had struck a nerve with practically everyone.

"Let's start with a simple question," Gary began. "What tagging standards do your teams currently use?"

The answers confirmed what he'd already seen in the data.

"We tag everything with application, environment, cost center, and owner," said Peter.

"We use project and department tags," offered a developer from Retail Banking.

"Our team doesn't require tags," admitted a lead from Wealth Management. "It's optional based on developer preference."

Barry's representative from Commercial Banking spoke up confidently. "We've been using a formal tagging schema for two years. Application, environment, service tier, backup frequency, data classification, and eight other required fields."

Gary paused. "Okay... but I have to say, I saw some pretty nonstandard tags." He glanced at his screen. "'BobTest2'? 'idk-ask-jenny'?"

The Commercial Banking representative flushed. "Well, compliance isn't 100%, but the standard exists."

The room erupted in a mix of defensive explanations and complaints about the difficulty of enforcing standards. Everyone seemed to agree that tagging was important, but nobody agreed on what tags to use or how to make people follow the rules.

"Hang on," Gary interrupted, cutting through the noise. "Let's take a step back. Before we talk about what our standard should be, let's agree on why we need one."

"Cost allocation," Fran said immediately.

"Security classification," added Sarah.

"Disaster recovery prioritization," offered Oscar.

The list grew: compliance requirements, support routing, automation triggers, resource organization. Everyone had a reason they needed tags, and they were all valid.

"So the challenge," Gary summarized, "isn't just agreeing on what tags to require, it's balancing completeness with usability. Too few tags, and we don't have enough information. Too many, and people won't fill them out correctly."

Gary had anticipated this deadlock. He'd seen how these discussions typically spiral into complexity, with each team advocating for their specific needs until agreement becomes impossible.

He pulled up a slide he'd readied before the meeting.

"What if we started with just three mandatory tags: 'environment,' 'owner,' and 'application'? Everything else would be recommended but optional."

It was a deliberate simplification, the minimum viable standard that could still provide meaningful context while being simple enough that teams might actually implement it.

Only three?" Barry's representative objected. "That's not nearly enough. Our business is far more complex."

"Three consistent tags are more valuable than fifteen inconsistent ones," Gary countered. "We can add complexity once teams adopt the basics."

The debate continued for another hour. Everyone agreed standardization was necessary, but they couldn't agree on what the standard should be. Each division had different priorities, different workflows, and different levels of maturity in their tagging practices.

"I think we need to take this offline, back to our teams," the Retail Banking representative finally said. "Get broader input before we decide."

Gary noticed the pattern: another delay, another round of consultations, another postponement of actual action. But unlike the cost optimization initiative, he couldn't easily demonstrate the value of standardized tagging in a two-week pilot. This was foundational work that would require broader buy-in.

"Alright," he conceded. "Let's take it offline. But let's collect feedback over email and reconvene next week with the goal to agree on an approach. And meanwhile, could everyone share their current tagging schemas so I can help identify common patterns?"

As the room emptied, Peter lingered. "They're going to come back with a dozen different requirements. Each team thinks their use case is special."

"Probably," Gary admitted. "But I had to try the collaborative approach first."

"The trick is finding a baseline everyone needs. The rest can come later."

Later that day, Gary received a message from Charles asking for an update on the tagging initiative. After the town hall had gone sideways, Charles was obviously keeping a closer eye on Gary's projects.

They met briefly in Charles's office.

"So, how's it going, Gary?" Charles asked, settling behind his desk.

"I think I understand the problem better now," Gary replied. "Everyone agrees tagging is important, but nobody wants to change their current approach. And every division needs something slightly different."

"So, what's your plan?"

Gary considered his options. After the town hall disaster, he was wary of pushing too hard. But without some level of standardization, every other initiative would be hampered by a lack of context.

"I'm going to try one more collaborative approach," he decided. "If that doesn't work... then I might need your help mandating a minimal standard that everyone has to follow."

Charles nodded thoughtfully. "Let me know when you get to that point. Just make sure you've exhausted the cooperative options first."

Day 83: Peter's Progress

Peter watched yesterday's tagging standards meeting unfold with a mix of frustration and déjà vu. Another well-intentioned initiative that seemed simple in concept but proved complex in practice, bogged down by competing divisional requirements and organizational complexity.

Gary's frustration was visible as the room emptied. Three simple, mandatory tags seemed like such an obvious starting point. But to the other divisions, it was either too simplistic or too intrusive. There was no middle ground to be found.

When Peter showed Gary how his team had been handling tagging for the past two years, he looked like he wanted to bang his head against the wall. They'd kept it straightforward initially: application, environment, cost center, and owner. All automated through their deployment pipeline so developers didn't have to think about it. Over time they added many more context tags, but it was only through iteration, without compromising the ease of automation.

"Why is this so difficult for the other teams?" Gary asked him after the meeting.

Peter didn't have a simple answer for him. The truth was complicated. A mix of team history, technical debt, and organizational culture.

His team had been fortunate. When they rebuilt the payments platform, they had the opportunity to establish modern practices from the ground up. They invested in building an in-house development team with the right skills instead of relying on contractors. They established a DevOps mindset early, breaking down the traditional silos between development and operations.

For most other divisions, it was a different story. They were carrying legacy applications, working with distributed teams, and often lacked the engineering talent to implement automation. What seemed simple to them, like automated tagging through a development CI/CD pipeline, was a major undertaking for them.

It wasn't that they didn't want to improve. It was that the gap between where they were and where they needed to be felt insurmountable. Gary thanked Peter for taking the time to talk with him, and they headed out of the conference room.

Later that day, Dennis stopped by Peter's office. He'd been at the tagging meeting and witnessed the exchange between Peter and Gary.

"Your team makes this all look so easy," he said, leaning against Peter's doorframe. "How did you find the time to prioritize quality of life improvements and backend operational features? We're always so buried in the next customer-facing deliverable."

"That's the paradox," Peter explained. "We actually prioritized those foundational elements early on, even when it meant delivering customer features more slowly at first. The tagging system is built into our CI/CD pipeline. If the required metadata isn't provided, the pipeline fails."

"And your leadership was okay with slowing down feature delivery?" Dennis asked incredulously.

"It wasn't an easy sell," Peter admitted. "But we demonstrated how these investments would accelerate everything downstream. We did a small proof of concept first. We automated one operational task and showed the time savings. Then we reinvested those savings into the next automation."

Dennis nodded slowly. "So, you built a virtuous cycle."

"Exactly. The initial cost is high, but the compounding returns are enormous. Now we deploy faster, troubleshoot quicker, and spend far less time on operational overhead."

Peter had seen a certain cycle repeated across Goliath for years, and he'd worked hard to avoid it. Teams couldn't improve their processes because they were too busy delivering features. But they were constantly slowed down by manual processes, which left less time for improvements. A perfect catch-22.

"Look, I know how hard it is to break that cycle," Peter told Dennis.

They sat in silence for a moment as Dennis seemed to contemplate the mountain of technical debt his team was facing.

"You know," Peter finally said, "I'm happy to offer advice whenever you need it. And there are others in the organization who want to help as well." He paused, considering who else could be a good soundboard. "That new cloud guy, Gary, he seems genuinely interested in making a difference. Have you worked with him on any of his initiatives?"

"Not yet," Dennis admitted. "Though his cost optimization idea did catch my attention during the architecture review."

"That should be an easy win," Peter suggested. "It's minimally invasive and could save your team significant cleanup time. Those hours could then be invested in your next automation."

Dennis considered this. "You make a good point. I'll talk to my team about prioritizing it. Maybe we can start small with just the cost script and see what resources it frees up."

As he left, Peter felt the familiar mix of satisfaction and frustration. On one hand, it was gratifying to help another team see the value of better practices. On the other hand, he was tired of having these same conversations year after year.

For a long time, Peter had questioned whether he was at the right company. Other organizations might offer environments where his team's approach would be the norm, not the exception. Places where he wouldn't always be the first one asked to pilot every new initiative.

But there was also something rewarding about being the agent of change, about showing what was possible within Goliath's constraints. And occasionally, like with Dennis today, he saw small shifts in thinking that might eventually lead to broader transformation.

The debates over standards and best practices would continue. The resistance to change would persist. But maybe, team by team, conversation by conversation, they could gradually raise the bar for how they operated in the cloud.

In the meantime, his team would keep delivering, keep innovating, and keep demonstrating that there was a better way to work, even within the constraints of a century-old bank. And if Gary's initiatives helped other teams start moving in that direction, all the better. They could use some allies in their quest to modernize Goliath.

Day 90: Endless Loop

Gary stared at his inbox with growing frustration. Over a week had passed since he'd sent his revised tagging proposal to the divisional representatives, incorporating their feedback from the meeting. So far, he'd received exactly two responses.

One from Peter's team: "Looks good. Ready to support implementation whenever others are on board."

And another from Wealth Management: "Need more time to review. Will share feedback by next week."

The rest? Silence.

"Any update on the tagging initiative?" Fran asked, stopping by his desk.

"Still waiting on responses," Gary replied. "I've followed up twice, but most teams aren't engaging, and getting a follow-up meeting took forever to schedule."

"Classic Goliath delay tactic," Fran said. "If you don't want to say no, just don't respond at all. Just wait until the next meeting to re-hash the same discussion."

Gary pulled up his project tracker. The cost optimization initiative that had started with such promise was now stalled across most divisions. The tagging standards had devolved into a never-ending review cycle. Even the basic security recommendations he'd shared were sitting in what felt like permanent "consideration."

A notification popped up in the corner of his screen: "New email from Barry: 'Per our division leadership meeting, Commercial Banking will be proceeding with...'"

For a moment, Gary felt a spark of hope. Had Barry's team finally come around? He clicked to open the full message.

"Per our division leadership meeting, Commercial Banking will be proceeding with our own tagging initiative based on our existing standards. We appreciate central IT's input but have determined our approach better meets our specific requirements."

Of course they had.

The pattern was becoming painfully obvious. Every attempt at standardization was met with delays, questions, and alternate proposals, and it was going to be too difficult to rally all the teams together for one agreed-upon approach. And tagging was just one of many initiatives, but it tied into other underlying issues such as security gaps, wasted resources, and operational confusion, all of which continued to grow.

It was just yesterday when Sarah came over with yet another security report showing vulnerabilities across multiple accounts. "It would be a lot easier to address if we had consistent tagging and ownership information," she noted, echoing Oscar's frustration.

"I know," Gary replied. "But we can't seem to get consensus on even the most basic standards."

"Welcome to my world," Sarah said. "I've been trying to implement basic security controls for two years. Everyone agrees they're important in principle, but no one wants to be the first to implement them."

For the first time since joining Goliath, Gary found himself questioning whether the collaborative approach was even possible in this environment. Peter's team had proven what was possible with the right practices, but they seemed to be the exception rather than the rule.

Maybe it was time for a more direct approach. If the divisions wouldn't align around common standards voluntarily, perhaps they needed stronger guidance from the top. Not as a permanent solution, but as a starting point to break the endless cycle of discussion without action. This worked for him at his last company, he took initiative to get things done and others came along for the ride. When something was wrong, they quickly iterated. The alternative was to be trapped in a constant planning cycle, and they simply didn't have time for that.

Gary opened a new document and began outlining a more comprehensive framework, not just suggested best practices, but specific standards with implementation timelines. It was a significant departure from his initial approach, but the status quo was failing.

"Three months in and still chasing basic standards," he muttered to himself. "Something has to change."

Day 97: Breaking Point

Gary stared at the latest cloud cost report, his frustration mounting. Despite the success with Peter's team and a handful of others, the larger divisions continued to resist even the most basic optimizations. The wasted spend was now projected to exceed $1 million annually, money that could be funding innovation instead of idle resources.

"Three months in and all I have to show is a few pilot projects," Gary muttered to himself. The early wins with friendly teams had created momentum, but that momentum had crashed into the wall of divisional politics.

Three separate tabs were open on his browser: a presentation deck from Commercial Banking showing their "cloud optimization strategy" with no actual implementation timeline, a security report from Sarah highlighting critical misconfigurations across multiple accounts, and a spreadsheet tracking the fifteen different tagging schemas currently in use.

His phone buzzed with a message from Fran: "CFO asking about cloud cost projections for next quarter. What should I tell him?"

Gary leaned back in his chair. The weight of the situation had settled on him over the last few days.

Collaboration and gentle persuasion weren't working. The divisions listened politely, acknowledged the problems, then continued doing exactly what they'd always done.

He closed his laptop and made a decision. Charles had mentioned an "open door policy" during his onboarding. Time to take that literally.

Twenty minutes later, Gary was standing outside the Head of Technology Operations' office, wondering if he was making a mistake. Taking problems to leadership without solutions wasn't his style, but he was running out of options.

"Gary," Charles looked up from his desk, surprised. "Come in. How's it going? Cloudy?" he added with a small chuckle at his own joke.

"You could say that," Gary replied with a polite smile as he settled into a chair across from Charles's desk.

"What's on your mind?" Charles asked, setting aside the document he'd been reviewing.

"Honestly? I'm hitting roadblocks," Gary admitted. "We've proven the value of our approach with Peter's team and a few others, but the larger divisions are stonewalling. I've got meetings, I've got working groups, I've got pilots, but I don't have actual progress at scale."

Charles nodded, his expression neutral. "And what do you think is the barrier?"

"The consensus-driven culture," Gary said, finally voicing what had been frustrating him for weeks. "Everyone needs to agree before anything changes, which means nothing ever does. And for an IT organization, that's particularly strange. Technology changes constantly, but our approach to managing it seems frozen in time."

"Go on," Charles said.

"At my previous company, when we identified waste or security issues, we fixed them. Here, we form committees to discuss fixing them. The cloud doesn't wait for consensus, it keeps evolving, keeps consuming resources, keeps creating new risks while we're still debating how to address the old ones."

Gary pulled up a draft of the document he'd been working on, an outline of key cloud operational standards covering security, cost management, and resource organization.

"This isn't a comprehensive model," Gary explained. "But these are foundational practices that every team should follow. Basic security controls, simple cost optimizations, consistent tagging. The things that enable rather than restrict."

Charles studied the document. "This is quite a shift from your initial approach. The divisions won't like it."

"They don't like the current situation either," Gary pointed out. "Barry complains about security gaps, but resists security standards. Dennis wants better cost control, but won't implement proven optimizations. Everyone acknowledges the problem, but no one wants to change their behavior."

Charles was silent for a moment. "You know, we're a bank for a reason. We move slowly because mistakes can be catastrophic. The consensus-driven culture you're frustrated with, it's also what keeps us from making reckless decisions."

"I understand that," Gary said. "But there's a difference between thoughtful caution and organizational paralysis. Right now, we're closer to the latter."

Charles paused, then offered an alternative. "I've been thinking about getting broader visibility for the cloud challenges we're facing. What would you think about presenting at a town hall with all the division IT leaders and their teams? We can call it the "State of Cloud at Goliath" on the agenda. You could make your case directly about why these baseline standards matter."

"A town hall?" Gary hadn't expected this opportunity. "That would give us the visibility we need."

"But be prepared for pushback. And remember, in this organization, how you introduce change is often more important than what the change is."

After leaving Charles's office, Gary sat quietly at his desk, contemplating the path ahead. He'd wanted to build consensus, to bring teams along through collaboration and shared success. But the urgency of the situation demanded a more decisive approach.

He'd seen the problems. He had the data. He had an outline of what needed to be done. It was time to stop asking for permission and start driving real change.

Day 105: The Big Announcement

"This is a mistake," Sarah whispered as Gary adjusted his collar in the reflection of his laptop screen. "You're moving too fast."

"We've been moving too slow for months," Gary replied, reviewing his presentation slides one last time. "Sometimes you need to rip off the band-aid."

The quarterly IT town hall was already underway when Gary and Sarah slipped into the back of the large conference room. Charles had been running through the standard updates, security metrics, project status reports, and the usual departmental highlights. Most attendees were half-listening while checking emails or finishing reports due later that day.

Division IT leaders and their key team members filled the seats, with Barry from Commercial Banking and his four direct reports occupying the front row as usual. Dennis sat several rows back, focused on his laptop screen.

Gary's presentation was scheduled toward the end of the agenda. Charles had slotted him in just a week ago, after their conversation about taking a more direct approach to cloud standards. As the Finance team wrapped up their quarterly budget review, Gary gathered his materials and made his way to the front.

"Next up, Gary will share his assessment of our cloud operations and propose a path forward," Charles announced, yielding the podium.

Gary connected his laptop and jumped straight into the presentation. "Over the past three months, I've been analyzing our cloud environments across all divisions. I've met with many of you, reviewed your processes, and identified both strengths and opportunities. Today, I want to share what I've found and what we need to do about it."

He displayed the first slide, a simple graph showing cloud spending growth over the past two years, with projections for the next three.

"Our cloud spend is projected to triple by next year. That growth isn't inherently problematic, but only if that spending is efficient and aligned with business value. Currently, it's not."

Gary clicked to the next slide, showing examples of waste identified across various accounts. Unattached volumes, idle instances, over-provisioned resources, all adding up to millions in annual waste.

"We've proven with several teams that basic cost optimizations can reduce spending by 15–30% with zero impact on performance or reliability. Yet these optimizations aren't being implemented consistently."

He continued through the presentation, methodically laying out the challenges in security, operations, and resource management. The data was compelling, the examples specific, the case for change undeniable.

Then he reached the slide titled "Cloud Standards Framework," and the room's atmosphere subtly shifted.

"Based on these findings, we're implementing a new set of cloud standards effective next month," Gary announced. "These will apply to all divisions and all cloud accounts."

He detailed the key components: mandatory tagging standards, required security controls, automated cost optimizations, and standardized deployment patterns.

"Central Cloud Services (CCS) will provide the tools and support to implement these standards, but adherence will be mandatory. Going forward, all cloud resources must comply with these baseline requirements."

The shift in the room was immediate and palpable. Barry sat up straighter, his expression hardening. Dennis closed his laptop. Even some of the teams that had been working productively with Gary looked uncomfortable.

"These standards aren't meant to restrict innovation," Gary continued, sensing the tension. "They're designed to create a secure, efficient foundation everyone can build on. Think of them as guardrails, not gates."

As Gary finished his presentation and opened the floor for questions, Barry was the first to speak.

"Let me make sure I understand," he said, his voice measured but cold. "Central IT is now mandating how my division manages its cloud infrastructure? The same infrastructure we've been successfully running for years without these... standards?"

"The standards address gaps we've identified across all divisions," Gary responded. "Including significant security and cost optimization opportunities in Commercial Banking."

"Gaps according to whom?" Barry challenged. "My team has maintained five-nines availability while delivering business-critical features on schedule. Now you're telling us we need to change our processes based on three months of analysis?"

Before Gary could respond, the Retail Banking director jumped in. "My concern is timing. We're in the middle of a major platform upgrade. Implementing these standards now would derail our delivery timeline."

"And who's going to fund the work to implement these changes?" another voice called out. "Our budgets and roadmaps are already set for the quarter."

The questions continued, growing increasingly pointed. Division leaders who had been cordial in one-on-one meetings were now openly resistant, unified in their opposition to what they perceived as central IT overreach.

Charles stepped in, attempting to redirect the conversation. "The intention here is to create consistency and efficiency. We're not trying to disrupt ongoing work."

"With all due respect," Barry responded, "any mandate from central IT disrupts our work. My team would need to divert resources from delivering business value to implementing these standards, regardless of how they're positioned."

Gary tried again. "The standards aren't arbitrary. They're based on proven patterns that are already working in some of our teams. Peter's payments group has implemented many of these approaches with significant success."

"Peter's team is different," someone called out. "They have different requirements, different skills, different priorities."

As the discussion continued, the divisions found common ground in resistance. Teams that normally competed for resources were united in pushing back against central standards. Division leaders who rarely agreed on anything were nodding in unison at each other's objections.

Eventually, Charles thanked Gary and moved on to the next agenda item, a brief update on the data center consolidation project. As Gary returned to his seat, he could feel eyes on him from around the room. What he had intended as a rallying cry for improvement had instead become a lightning rod for division resistance.

The town hall continued with the remaining agenda items, but Gary barely heard them. He was replaying his presentation in his mind, wondering where it had gone wrong. The data had been solid. The case for change was compelling. The standards themselves were reasonable.

When the meeting finally concluded, Gary gathered his materials and headed back toward his office.

"That could have gone better," Charles said, catching up to him in the hallway.

Gary tried to mask his disappointment. "I knew there would be resistance, but I didn't expect a unified front against even discussing the standards."

"You presented it as a decision, not a conversation," Charles observed. "In an organization this size, with divisions this powerful, that approach rarely works. Even when the underlying ideas are sound."

Later in the day, Sarah dropped by, offering a sympathetic smile. "For what it's worth, you weren't wrong about the problems. Just about how ready the organization is to address them."

"I thought the waste and security issues would be compelling enough on their own," Gary admitted. "I underestimated how much these teams value their autonomy, even when it comes at a cost."

"At Goliath, how you introduce change is often more important than what the change is," Sarah echoed Charles's earlier advice. "The divisions need to feel ownership, not compliance."

Day 112: Barry's Victory

Twenty years at Goliath gave Barry perspective. He'd seen central IT initiatives come and go like quarterly fashion trends. Grand visions announced with fanfare, implemented half-heartedly, then quietly abandoned when the next big idea came along. Gary's cloud standards push was just the latest chapter in this predictable story.

Barry didn't set out to torpedo his initiative. The man had good intentions and some valid points about cost optimization. But the moment he tried to mandate how Commercial Banking ran its technology, the outcome was inevitable. His division generated a third of the bank's profits. They didn't take orders from support functions; they set their own course.

When Barry first joined Goliath, central IT controlled everything. Every server, every application, every change request had to navigate their bureaucracy. Projects that should have taken weeks stretched into months. Innovation stalled. Customers suffered.

That's why he built Commercial Banking IT differently. They seized control of their destiny, building their own development teams, establishing their own processes, moving to the cloud on their own terms. Some called it "shadow IT." Barry called it survival.

The results spoke for themselves. While other divisions waited for central approval, Commercial Banking IT delivered. Digital banking platforms that kept them ahead of FinTech startups. Mobile payment solutions that customers actually wanted to use. Analytics capabilities that transformed their lending decisions.

Barry still remembered when they first moved to AWS. Central IT insisted they wait for their "enterprise cloud strategy," which was still two years away from completion. Meanwhile, their competitors were already leveraging cloud elasticity to scale services during peak periods. They had to choose: innovate or fall behind.

They moved ahead. Set up their own cloud accounts, built their own security controls, established their own operating practices. They made mistakes, certainly. But they were their mistakes, quickly identified and fixed before they affected customers.

By the time central IT published its cloud strategy, Commercial Banking already had dozens of production services running in AWS. Central IT wanted them to migrate everything to their "approved architecture." Commercial Banking declined, and even threatened escalation. The department showed the business results that their approach had delivered.

That was the end of that discussion.

Walking into that meeting last week, Barry knew exactly what was coming. He'd read Gary's preliminary findings and even agreed with some of his observations. But then came the slide with 'Central Cloud Services (CCS)' plastered across the top. Yet another acronym in IT's endless alphabet soup announcing their next plan to 'help' them.

The room's reaction was priceless. Jennifer from Retail Banking actually rolled her eyes. Walter from Wealth Management started typing furiously on his phone, probably messaging his team about another central IT land grab. Even Dennis, who usually stayed quiet in these meetings, looked visibly annoyed.

When Barry spoke up, he wasn't just defending Commercial Banking, he was giving voice to what everyone was thinking. They didn't need central IT swooping in trying to manage their business. They had no context or understanding of how they operated.

Another central IT initiative crashing and burning came faster than he expected. In the hallway after the town hall, division heads gathered in impromptu clusters, united in opposition.

"Did that cloud guy really just try to mandate new standards on us?" Jennifer asked.

"Central Cloud Services," Walter repeated with a smirk. "As if we need another three-letter acronym."

Dennis surprised Barry with his directness. "I don't have time for more overhead, just going to keep my head down and get things done."

Even Charles seemed to sense the shifting winds. He called Barry yesterday, originally to discuss the quarterly architecture review, but his real purpose was evident.

"Barry, I could tell you were annoyed during the town hall," he began.

"Not annoyed, Charles. My division delivers. Consistently. Any mandate that threatens that delivery is a non-starter."

There was a pause on the line. "Your teams do consistently deliver, but we know there is always room for improvement. What if we approached this differently? What if each division could implement the standards in their own way, on their own timeline?"

"Then they wouldn't be standards, would they?" Barry replied. "They'd be recommendations. And we're always open to recommendations, especially when they're backed by data. What we're not open to is being forced to meet requirements my team doesn't have time for."

This morning, Barry had received the agenda for next month's IT leadership council. Gary's cloud standards had been reframed as "recommended best practices" for division consideration. A big victory, and a necessary one.

His team had already reviewed Gary's cost optimization findings. The so-called "discoveries" were hardly revelations to them. They had identified these optimization opportunities months ago and already had them on their roadmap. They'd implement them on their schedule, not because central IT suggested them, but because they aligned with plans they'd already developed. And when these optimizations showed results, it would be a Commercial Banking initiative, another example of how they led while others followed.

They were all on the same team, but the business units led the way. When would their leaders understand that a decentralized model worked best for everyone?

Day 120: The Aftermath

Two weeks after the town hall, Gary could still feel the ripple effects. Meeting invitations went unanswered. Emails received delayed responses. Teams that had been working productively with his group suddenly had "competing priorities."

"This morning, I got copied on a message from Barry to his team," Fran said, stopping by Gary's desk. "They're implementing their own cost optimization initiative. Using a lot of the same approaches we presented, just with a different name."

"And without collaborating with us," Gary added.

"Without even acknowledging us," Fran corrected. "The message made it sound like it was their idea from the start."

Gary wasn't surprised. It was the third time this week he'd heard about a division implementing pieces of his standards framework while carefully avoiding any mention of central IT involvement. Retail Banking had started a tagging project. Wealth Management was reviewing S3 bucket permissions. Each division was working independently, duplicating efforts and reinventing the wheel.

"The data center migration planning is seeing similar resistance," Fran noted. "Every division has suddenly decided they need to revisit their migration plans independently."

Gary nodded. The town hall hadn't just stalled his cloud standards. It had reinforced the walls between central IT and the divisions. By pushing too hard, too fast, he had inadvertently strengthened the very silos he had been trying to break down.

His phone buzzed with a calendar reminder: "1:1 with Charles - 30 min."

As Gary headed to Charles's office, Sarah caught up with him in the hallway.

"Heard you're meeting with Charles," she said. "A heads up, Isabel from Internal Audit is actually interested in your framework. She told me they've been struggling to create specific controls for cloud services. Their traditional frameworks were designed for on-premises systems."

"That's something, at least," Gary replied.

Charles was on the phone when Gary arrived, gesturing for him to take a seat. The conversation appeared to be about delayed projects and missed expectations, likely fallout from the increased divisional resistance.

"Sorry about that," Charles said, hanging up. "Operations update from Wealth Management. They've hit some roadblocks with their cloud migration."

"I've been thinking about next steps after the town hall," Gary said. "I misjudged how the approach would be received."

Charles nodded. "The underlying issues still exist, such as resource waste, security gaps, and inconsistent operations. But I've noticed the divisions are starting to address some of these on their own terms."

"Yes, but without coordination," Gary noted. "Each division is implementing its own version of the same solutions."

"At least it's sparked some action," Charles replied. "Though I am concerned about the duplicate effort and that their momentum might fizzle when other priorities emerge."

"Isabel from Audit is apparently interested in the framework," Gary mentioned. "Sarah just told me they've been looking for cloud-specific controls."

"That could be useful," Charles acknowledged. "Compliance requirements tend to get attention."

He turned his attention to Gary. "So what's next, Gary? Where do we go from here?"

The question caught Gary off guard. He'd expected criticism, not an invitation to chart the path forward.

"I think we need to focus on where we can add value without threatening autonomy," Gary said after a moment. "Find the areas where divisions actually want help and start there. Build credibility through collaboration rather than direction."

"That sounds like a better approach," Charles agreed. "Specific examples?"

"Shared tools that divisions can opt into. Common patterns they can implement their way. Training and enablement rather than enforcement." Gary paused. "And maybe we focus first on new accounts and workloads, where there's less resistance to change."

"Good," Charles said. "And remember, this is a marathon, not a sprint. The organization didn't develop these patterns overnight, and they won't change overnight either."

Back at his desk, Gary felt a bit defeated. The past two weeks had been a difficult lesson in organizational dynamics. Technical solutions weren't enough when the real challenges were cultural and political.

He opened a new document and began drafting a different approach. Not another standards framework, but a plan for rebuilding relationships with the divisions. What could his team offer that divisions would actually want? Where could they add value without threatening autonomy? How could they enable teams to move faster with less effort?

Day 127: Warning Signs

"You need to see this."

Sarah placed a printed report on Gary's desk, her expression grave. It was a security scan of the Commercial Banking division's development accounts. Red highlights and critical warnings filled the pages.

"These are just from the last two weeks?" Gary asked, flipping through the report.

"Yes, and it's getting worse," Sarah replied. "We've gone from twelve critical findings last month to thirty-seven this month. S3 buckets with public access, IAM roles with admin privileges, missing encryption, API endpoints without proper authentication."

Gary studied the report more closely. Most of the vulnerabilities were in development environments, but a concerning number had appeared in staging accounts that sometimes contained production-like data.

"Barry's accounts are particularly concerning," Sarah continued. "While they handle the most critical issues eventually, there's a growing backlog of security findings. It's the third report we've sent them this month."

"Have they responded to any of the notifications?"

Sarah nodded. "They acknowledge the tickets and fix the most glaring issues, but the majority never reach their priority threshold. They're too busy delivering features to address what they consider 'theoretical' risks. And it's not just Commercial Banking. Everyone's rushing to deliver application features. Security is an afterthought."

"And you're still actively notifying teams when you find issues?" Gary asked.

Sarah hesitated. "Yes, but... less forcefully than before. Given recent tensions since the town hall, I've been treading carefully with how we communicate findings."

This was concerning. The divisions were pushing back against any perceived interference, even when it came to legitimate security concerns.

"What happens if one of these vulnerabilities gets exploited?" Gary asked.

"Best case? Unauthorized access to development resources. Worst case? Data breach, especially if they're using real customer data in those environments," Sarah's voice was tight. "And if regulators find out we identified these risks but didn't address them..."

"We'll be in violation of multiple policies," Gary finished.

The irony wasn't lost on him. The very standards he'd proposed would have prevented most of these vulnerabilities from being created in the first place.

"We need to get ahead of this," Gary said, turning to his laptop. "I've been working on a new approach, less about standards, more about shared tools and enablement."

He pulled up the document he'd been developing over the past week, a framework of cloud enablement services. Self-service security scanning, automated cost optimization, template-based deployments with best practices built in.

"This is good," Sarah said, reviewing the document. "But will it address the immediate risks? These vulnerabilities exist today, not tomorrow."

"No," Gary admitted. "This is the long game. For the immediate issues, we need a different approach."

They spent the next hour developing a strategy. Instead of sending yet another notification that would be ignored, they would create dashboards showing each division's security posture compared to their peers, appealing to competitive instincts rather than compliance requirements. They would target the most critical vulnerabilities with specific remediation guidance, not just problem statements.

"I'm also going to talk with Isabel," Gary decided. "Not to escalate, but to understand the regulatory requirements better. The divisions might ignore us, but they can't ignore compliance mandates."

As Sarah left, Gary stared at the security report again. The warning signs were flashing red, but the organizational dynamics made addressing them incredibly complex. It wasn't just about technical solutions anymore. It was about finding the right pressure points in a resistant system.

The vulnerabilities in Barry's accounts were particularly troubling. Thirty-seven critical findings, all acknowledged, none prioritized. How long before one of those theoretical risks became a real incident? The divisions couldn't keep ignoring these warnings forever.

Day 134: First Alert

Sarah had spent months getting her CNAPP tool access to Commercial Banking's most critical cloud production accounts. Since then, the backlog of critical security issues identified had been growing, and none of the alerts had received any response from application teams.

She called Gary. He picked up right away.

"Gary, remember that vulnerable IAM role we flagged in Dennis's FinTech environment with overly permissive S3 permissions? Our CNAPP tool just triggered a high-priority alert on suspicious activity patterns using that role."

"What kind of suspicious activity?" Gary asked, already reaching for his jacket.

“The CloudTrail audit logs from AWS are showing that role is being used from outside our corporate network. The access pattern is highly abnormal with systematic bucket enumeration followed by high-volume GetObject calls that indicate someone is downloading large amounts of data.”

While her team had identified numerous vulnerabilities over the past months, this was different. It wasn't just a potential risk but an active threat unfolding in real time. Sarah's voice had a tension Gary hadn't heard before.

The shift from theoretical to immediate danger had put her entire team on edge, and they were scrutinizing the alert patterns with heightened vigilance.

"Why the urgency on this one?" Gary asked.

"This isn't just another vulnerability. It's active exploitation. With all the findings we've been reporting getting ignored..." Sarah took a breath. "And I believe it's accessing customer financial records. If this is an actual breach, it would be regulatory reportable."

Gary's pulse quickened. "Have you notified the CISO?"

"I had to. She's pulling together an ad hoc incident response team and has notified Charles. I've also sent an urgent message to Dennis's team, but haven't heard back."

Gary called Dennis directly. After several attempts, he finally got through.

"I'm in a client presentation," Dennis whispered. "Can this wait?"

"No, it cannot," Gary replied firmly. "We have an active security incident in one of your cloud accounts. Someone is accessing customer data in S3 buckets using a monitoring service role from outside our corporate network."

"What? How would you know? Are you sure?"

"Sarah's security monitoring detected unusual API call patterns. We need your team in the security conference room immediately for incident response."

"I'll be there in twenty minutes. Mike, my technical lead, will meet you right away."

After a few minutes, multiple teams were alerted. The conference room was filling, forming an incident response. Sarah was displaying the alerts and related log entries to the room, describing the situation as the CISO, Cathy, walked in.

"This IAM role has permission to read S3 bucket contents," Sarah explained. "Someone is using it from outside our network to systematically download customer records."

Mike, squinting at the screen, decided to look up the source IP address common across the logs. "The address is in our metro area, but that doesn't tell us much."

"Could it be an employee working from home?" Gary suggested.

"It's possible," Mike frowned, "but we have strict protocols about handling customer data. This kind of bulk access would require a change ticket."

Cathy was visibly concerned. "This pattern matches what we've seen in previous compromise scenarios. Broad enumeration followed by targeted data access. The fact that it's coming from outside our corporate network is particularly worrying."

"What containment options do we have?" Gary asked.

"The quickest option is to update the IAM policy to remove the GetObject permission," Sarah replied. "That would block the data access while we investigate."

"Would that impact production?" Gary asked Mike.

"The role being used is for our monitoring service which only needs to read bucket configurations, not the actual data. It never needed GetObject permissions in the first place."

"Let's do it," Cathy ordered. "And I want a Microsoft Teams bridge opened immediately for the entire application team. We need to determine if anyone authorized this access."

As Sarah implemented the containment measure, Dennis arrived with Barry. "What's happening?" he asked, looking concerned.

While Gary briefed them both, Sarah's team set up the conference bridge. They broadcast the meeting link to Dennis's entire team through their emergency notification channel.

Within moments, a developer named Jason joined the call.

"Hello? This is Jason. I got the emergency notification to join this bridge."

"Jason, this is Mike and we have a number of others in the room. We're investigating unauthorized access to customer data from outside our network. Do you know anything about this?"

"Uh, that may be me," Jason replied, uncertainty in his voice. "I'm doing the data migration we discussed last week. Moving the verification records to the new storage layout."

The room fell silent.

Mike looked puzzled. "Jason, we didn't finalize that migration plan. There's no approved change ticket for this."

"You said it was urgent before month-end processing," Jason replied. "I couldn't come to the office because my kid is sick, so I decided to do it from home. I've done maintenance like this before without issues."

Cathy's expression grew serious. "Let me be clear, this is extremely concerning. You're accessing sensitive customer financial data from your home network, without proper authorization, without change control, and without using secure channels. This violates our security policies on multiple levels."

"Today exposed critical gaps that we need to address before a malicious actor exploits them," she concluded.

After the call ended, Barry stepped in. "So no actual security breach. Just a developer trying to be helpful."

"I wouldn't downplay this," Cathy warned. "A developer accessed and transferred massive amounts of sensitive customer data without proper controls. The only difference between this and a malicious breach is intent."

"And it's worth noting," Sarah added, "that we only detected this because we recently deployed our security tools to this particular account. We still do not have visibility into all our cloud accounts or a comprehensive understanding of who has access to what."

"The good news," Gary offered, recognizing the opening, "is that we can address these issues systematically. We need proper IAM permission boundaries, comprehensive account inventory, and resource tagging."

Barry, taken aback by the quick turn of conversation, had no option to object. "Let's discuss this further next week."

As the meeting concluded, Cathy outlined immediate actions: "First, we need a complete inventory of all cloud accounts and their critical resources. Second, we need to identify similar IAM vulnerabilities across our environment. Third, we need tighter change control processes. And fourth, mandatory VPN usage for all cloud administrative access."

"Today exposed critical gaps that we need to address before a malicious actor exploits them," he concluded. "I'll make sure to meet with Charles and the division leaders on this tomorrow. We need to strengthen our controls across the organization."

Day 137: Blind Spots

"So do we know how many other accounts might have this vulnerability?"

The question came from Charles, who had called an emergency meeting with Gary, Sarah, and the division IT leaders. Three days after Jason's unauthorized data migration had triggered their security response, the initial relief had given way to a deeper concern: if this could happen in one account, where else could it happen?

Gary glanced at the presentation he'd prepared. "That's precisely the challenge we're facing. When we tried to compile a comprehensive inventory of similar IAM roles across all divisions, we hit a wall."

Barry looked up from his notes. "What kind of wall?"

"Remember the tagging standards I proposed a while ago?" Gary asked. "One of the primary benefits would have been direct ownership and visibility across all cloud resources. Without consistent tagging and inventory, we're essentially searching in the dark."

Sarah nodded in agreement. "Even with perfect tagging inside the accounts we know about, we'd still have blind spots in accounts we haven't identified."

"We can only analyze what we can see," she continued. "My security tools can detect vulnerabilities, but only in accounts where they're deployed."

Gary added, "In hindsight, to the other initiatives I was advocating for, I see now that we should have focused first on just getting visibility and starting with what cloud accounts we have out there."

"We can't secure what we can't see. We can't manage what we don't know exists," said Charles.

Dennis, still visibly uncomfortable after the Jason incident, spoke up. "My team has been working with Adam from Asset Management and Sarah to identify all our accounts. We've found three that weren't in the official inventory."

Gary quickly commented, "We're starting to identify these accounts now and appreciate all the division leaders' cooperation."

"Glad we have an immediate path, but how do we make sure this doesn't happen moving forward, since we have no central record of our cloud accounts?" Charles asked.

"For the inventory," Gary continued, "Adam has already suggested working directly with Amazon, Microsoft, and Google to identify all accounts using our domain name for email contacts. We can also work with Finance to reconcile credit card statements against known and unknown accounts."

"And eventually," Gary added, "moving from individual credit card billing to centralized accounts would give us much better visibility and control. But that's a longer-term solution."

Barry, who had been unusually quiet, finally spoke. "I spoke with Cathy after the incident. She mentioned that Isabel from Internal Audit had reached out."

That caught everyone's attention.

"What about?" Jennifer asked.

"After hearing about the incident, she realized Audit hasn't adequately assessed our cloud landscape," Barry explained. "They've been focused on traditional infrastructure controls, but our growing cloud footprint isn't well represented in their scope."

Gary hadn't expected this development, but it made sense. Auditors would naturally want to understand the rapidly expanding cloud usage.

"Is this a formal audit?" Charles asked.

"Not yet," Barry replied. "Isabel wants preliminary discussions to understand what we're running in the cloud, what business purposes it serves, and how we're applying internal controls. She's looking to include cloud in their regular assessment scope going forward."

Barry paused. "The problem is, as we've just discussed, we don't have a consistent way to explain what we have in the cloud. Each team would give different answers, and that fragmentation will make us look disorganized. I'm not confident these discussions will go well."

The meeting shifted from feeling like they were getting ahead gradually to now having to take immediate action. Charles turned to Gary.

"You've been advocating for better cloud coordination since you arrived. What do we need to do right now to get ahead of this?"

Gary hadn't expected such a direct question, but he was prepared. "We need to accelerate our account inventory work. First, we identify all accounts. Then we establish ownership. Finally, we implement basic visibility across them."

"How long will that take?" Charles asked.

Gary looked at Sarah, who shook her head slightly. "With our current approach? Weeks, maybe months. We'd need to work with each division separately, navigate different processes, get multiple approvals."

"We don't have months," Barry stated. "Audit won't wait."

Gary saw his opening. "There is another way. If we coordinated efforts across divisions rather than working in silos, we could move much faster."

Dennis nodded. "My team is already working with Adam. We could expand that collaboration."

Charles looked around the room. "I want all divisions to prioritize the cloud account inventory effort. Gary will coordinate, but each of you needs to assign resources to help. Consider this a compliance priority."

There were nods around the table. Even Barry didn't object.

"I want weekly progress reports," Charles added. "Let's focus on getting a complete inventory first. Once we know what we have, we can better address the security vulnerabilities and prepare for Isabel's assessment."

As the meeting dispersed, Barry stopped Gary at the door. "I'll have Jennifer coordinate with your team on our account discovery," he said, surprising Gary with the direct offer of help.

As Gary left the room, he could hardly believe the shift. After months of pushing against resistance, he suddenly had cross-divisional cooperation. The Jason incident, combined with audit pressure, had created an opening that hadn't existed before.

His phone buzzed with a message from Sarah: "Did Barry just voluntarily offer help?"

Gary smiled as he replied: "Yes. Now we need to make the most of it before the momentum fades."

Day 140: Sarah's Crisis

Three years at Goliath, and Sarah had never seen anything like the past week.

Her desk was covered with printouts of security findings collected over months. Vulnerable IAM roles, exposed S3 buckets, excessive permissions. Warnings that went unheeded, tickets closed without action.

Now, suddenly, everyone cared.

The Jason incident changed everything. Not because it was severe, but because it exposed all their vulnerabilities at once. Functionally, it was a stress test revealing how fragile their cloud security really was.

Sarah stared at the list of accounts they were trying to inventory. Some they'd known about but couldn't scan. Others visible in billing but inaccessible. Each one potentially riddled with misconfigurations. And they still didn't know how many more existed.

Her security tools were good, but they were scanning only a fraction of their environment. She'd had to fight for every account connection. "Why does Security need access?" "We handle our own security." "That tool might impact performance." Every excuse in the book.

The CISO, Cathy, had been supportive, but couldn't overcome divisional barriers. Until now.

Gary stopped by, coffee in hand. "How's the account inventory coming along?"

"Better than expected. I've got deployment rights to those 27 new accounts. Even Barry's team is cooperating."

"Never let a good crisis go to waste," he said with a slight smile.

When Gary first arrived, Sarah wasn't sure what to make of him. Another central IT person with big ideas. But he got it. He was fighting the same battles, just from a different angle.

She gestured to her dashboard full of alerts. "Every new account we identify just adds to our workload. We're discovering issues faster than we can fix them. Hundreds of new findings this week alone."

"That's why we need to approach this systematically," Gary replied. "Complete the inventory first, then identify the critical systems, then prioritize what to fix."

"So we'll know exactly what was vulnerable when it's too late to fix it?" Sarah asked with frustration.

"We need to do both, but in the right order," Gary explained. "Think of it like a building on fire. First, you need to know the layout of the building and where people are trapped. Then you can prioritize your rescue efforts. Without the layout, you might waste time in empty rooms while people need help elsewhere."

Gary was right, and Sarah could see it. The account inventory effort had pulled in people from across the organization. Oscar from Operations, Adam from Asset Management, even Mike from Dennis's team. For the first time, they were working together rather than around each other.

But they were racing against time. They'd been identifying critical vulnerabilities for months: exposed S3 buckets, excessive permissions, insecure configurations. They reported them, teams acknowledged them, yet nothing changed.

"What keeps you up at night?" Gary asked.

"That we're focusing on account inventory while known critical issues remain unaddressed," Sarah admitted. "It feels like we're moving backward instead of fixing what we already know is broken."

Gary considered this. "Without a foundation of visibility and process, we'll just be playing whack-a-mole with alerts forever. We alert on issues, but that doesn't mean teams will act on them."

"So we ignore the fires we can see?" Sarah injected.

Gary replied. "No, we build systems to prevent fires in the first place. We're finally getting momentum to build a foundation where we can layer in controls upfront, not just react afterward. It feels slow, but this is the fastest this organization has moved since I arrived."

Sarah hadn't considered it that way. Maybe building the foundation first made sense, even with the alarms blaring.

Her phone buzzed with a message. Cathy needed her to join Barry to discuss their approach for the upcoming conversations with Isabel.

As she walked to the meeting, Sarah reflected on how much had changed. Just a week ago, Barry and Cathy were working around each other. Now they were aligned, both focused on the same priority.

The task ahead was enormous. They needed to map accounts, understand dependencies, identify critical systems, and prioritize remediation while continuing their day-to-day security operations. And she knew that once they started mapping the connections between accounts, they'd discover even more complexity than they expected.

Sarah couldn't help thinking that maybe the Jason incident was a good thing after all. Sometimes it took a near-miss to open people's eyes to the risks hiding in plain sight.

Day 143: Hidden Connections

It was day six of their account inventory initiative following the Jason incident. After Cathy's directive to identify all cloud accounts and their critical resources, Gary had been working out of a conference room that Charles' executive assistant had arranged for him. Sarah had been devoting significant time to help Gary, recognizing that this initiative directly addressed security vulnerabilities her team had been flagging for months. The room had quickly become cluttered with diagrams and notes. Whiteboards now covered every wall with account diagrams, IAM role mappings, and a growing list of security findings that seemed to multiply by the hour.

"Wait, so what you're saying is that your authentication service is calling our payment API, which is then accessing their customer database?" Gary asked, pointing to representatives from three different divisions.

"That's why we can't just shut down those cross-account access paths or revoke those IAM permissions," Mike from Dennis's team explained. "The entire mobile banking experience depends on these connections. You fix one security gap and you might break core functionality."

"This is exactly the problem," Sarah said. "We identified the broad permissions as security risks, but without understanding these hidden dependencies, we can't safely remediate them."

Gary studied the complex web of inter-account connections they'd mapped. What had started as a straightforward inventory to address the IAM vulnerabilities had revealed a far more complex ecosystem than anyone anticipated.

"So let me understand this," Gary said, turning to Jennifer's representative from Retail Banking. "Your team created this authentication service three years ago, but ownership transferred to Commercial Banking during last year's reorganization?"

"Only partially," the developer explained. "We still manage the infrastructure, but their team owns the application code. The service runs in our account but uses their IAM roles for access. Nobody documented this when responsibilities shifted."

Sarah nodded. "And that explains why our security scans flagged conflicting issues. We're seeing permissions that make no sense until you understand these cross-account dependencies."

Oscar spoke up. "We've been pushing for better dependency documentation for years. In an incident, we can't afford to be discovering these connections from scratch."

Gary checked the "Critical Issues" board, where they'd been tracking security gaps. The number had grown from 37 to 128 in just six days as they gained visibility into more accounts.

"Look at this Lambda function we found this morning," Sarah said, bringing up a diagram. "It has an IAM role with access to every S3 bucket in the account, not just the one bucket it needs."

"That was for testing," the Retail Banking developer admitted. "When we moved to production, we never scoped down the permissions."

Gary's phone buzzed with a message from Charles: "Need status update for executive committee by end of day. How many critical vulnerabilities have we identified? When will they be fixed? Isabel expects preliminary information tomorrow."

"We need to prioritize," Gary announced. "We can't address everything before meeting with Isabel, and we still don't have a complete picture of our accounts and their relationships."

"But who decides what's critical?" asked someone from Wealth Management. "Our trading platform might not touch as many customers as mobile banking, but it moves millions in transactions."

"Let's organize our findings into three categories," Gary suggested, sketching on the whiteboard. "Quick Wins we can fix immediately, Complex Fixes that require coordination, and Strategic Improvements to prevent recurrence."

As they developed the categorization framework, Barry appeared unexpectedly in the doorway of the conference room. He surveyed the whiteboards with an unreadable expression.

"I've been getting updates from my team," he said. "Things look worse than we thought, but at least now we can see what we're dealing with."

Gary smiled with understanding. "Yeah... it usually looks messier once you turn on the lights."

Barry studied the prioritization framework. "For what it's worth, the cross-account issues you've identified? We've been struggling with those for years. Never had the right lens to see the full picture."

After everyone left, Gary reflected on their discoveries. The account inventory had revealed something far more complex than isolated security issues. They weren't just tracking accounts to satisfy Audit's initial inquiries. They were uncovering a hidden network of dependencies nobody fully understood. Until they mapped these connections, even the simplest security remediation could have unexpected consequences.

Day 146: Coming Together

Three days had passed since they'd discovered the complex cross-account dependencies. What had started as mapping cloud accounts had naturally expanded as they uncovered more issues needing immediate attention.

The conference room had transformed in the past week. Originally just a place to track accounts, it now served as a hub where people from different teams would stop by to work on specific problems they'd identified.

"Dennis, what progress have you and your team made with those IAM permissions we discussed?" Gary asked as Dennis arrived with his laptop.

"We've begun tackling the excess permissions issue from two angles," Dennis explained. "First, immediate remediation of the most egregious cases, roles with admin privileges that don't need them. We've identified and prepared changes for 12 roles so far, though most are still pending approval through change control." He added with a slight smile, "Following all proper procedures this time around."

He brought up a diagram on his laptop. "Second, we're developing this permission framework that categorizes roles based on their function. For each category, we're defining a standard permission set that follows least privilege principles."

"Any pushback from the application teams?" Gary asked.

"Surprisingly responsive," Dennis said. "When we showed them the actual permissions their applications were using versus what they had granted, most were shocked by the gap. No one wants unnecessary permissions once they understand the risk."

As the day progressed, Gary spoke with others who'd been working on different aspects of the problems they'd uncovered. Nancy from Networking, who had been quietly getting involved over the past few days, had begun mapping the cross-account connections and categorizing them by risk level. Dennis and Adam continued expanding their account mapping, discovering new connections with each account they added.

As Gary listened, he realized something remarkable was happening. These weren't just technical fixes, but collaborative solutions emerging from people who'd been working in silos for years. They were sharing expertise, challenging assumptions, and building common approaches.

"There's something else," Adam added. "We're starting to document these patterns in a shared repository. Not just what to fix, but why these issues occur and how to prevent them in the future."

"A knowledge base?" Gary asked.

"That's the direction we're heading," Adam replied. "It could eventually become a pattern library, standardized approaches to common cloud challenges that any division can use."

Later that day, Gary updated Charles on their progress.

"Honest assessment," Charles said. "How's this approach working?"

"It's mixed," Gary replied. "Dennis and his team have been fully engaged since they were directly affected by the Jason incident. They're making real progress. But other divisions showed interest immediately after the incident and then faded back to business as usual."

"What about the central teams?"

"Fran, Sarah and Adam have been great, but they're constantly getting pulled back into their day jobs. I managed to get Nancy from Networking to help today, but we're still lacking engagement from the Identity folks and other central IT Shared Service units."

Charles nodded. "Division involvement is always challenging unless it's a direct priority for them. And our central teams are stretched thin as it is."

"That's the core issue," Gary agreed. "This work is proving valuable, but it's competing with everyone's regular responsibilities. I could use dedicated cloud specialists to maintain momentum."

"I'm working on that," Charles said. "Headcount is always a battle, but this incident has given us some leverage. In the meantime, keep focusing on tangible wins. Nothing builds momentum like demonstrated success."

Day 160: Audit Pressure

"I think we're ready for tomorrow's follow-up meeting with Isabel," Gary said, reviewing the slides one last time with Sarah.

Two weeks of work had given them unprecedented visibility into their cloud landscape. They had identified 128 critical security issues, with Dennis's team addressing a few of the most urgent ones, but the vast majority remained in various stages of assessment.

"The metrics show what we've discovered," Sarah noted, "but Isabel emphasized that she expects more than just identification and ad hoc fixes."

"Exactly," Gary replied. "She's looking for a sustainable process for managing these environments long-term, not just a one-time response to the incident. That's the part we're still developing."

The next morning, Gary found himself in a conference room with Isabel and her team, along with Charles, Cathy, and Fran. Isabel had already reviewed their status report and came prepared with specific questions.

"I appreciate the progress you've made with identifying issues," Isabel began. "But I'm concerned about sustainability. You've gained visibility, but remediation is happening sporadically without consistent processes."

Gary nodded. "We've been focusing on building a complete inventory and understanding dependencies before making significant changes."

"That's a start, but it only addresses a small part of the problem," Isabel replied. "What about the bigger picture? How are you inventorying cloud resources across the organization? What's your request process for new resources? How are cloud identities tied back to our corporate systems?"

She placed a binder on the table. "Our Internal Asset Protection Policies aren't being applied consistently in cloud environments."

"You're right," Gary acknowledged. "We've been responding to the security incident, but there's a broader set of controls we need to map to our cloud environments."

"Who's responsible for ensuring cloud resources meet these requirements?" Isabel asked. "How do we verify compliance to our standards?"

"We need to establish a formal structure for cloud environments, just like we have for our data centers," Isabel continued.

This was the compliance hammer Gary had been expecting, but rather than feeling threatened, he saw an opportunity.

"I agree completely," Gary said. "But we can't just force our existing operational processes onto cloud environments. That would kill the agility these teams need." He paused. "Cloud is software-defined infrastructure. It needs software-defined operations which ensures automated processes that can keep pace with the technology."

Charles looked interested. "What are you suggesting?"

"We need a coordinated approach across Shared Services," Gary explained. "A committee of key central IT functions working together to develop consistent standards and practices."

"So a central committee to control cloud usage?" Isabel asked skeptically.

"Not to control, but to enable," Gary clarified. "It would include Security, Operations, Finance, Identity Management, Networking. Its purpose would be to develop consistent approaches that make it easier for divisions to do the right thing."

"Instead of reacting to different requirements for each new cloud account, we'd develop unified standards and automation to provision accounts properly from the start."

Charles nodded with growing enthusiasm. "This could solve multiple problems. Each central function developing standards in isolation has created confusion. A coordinated approach would be much more effective."

"How would this committee ensure adherence to our policies?" Isabel asked. "We need verifiable controls, not just standards."

"That's key," Gary replied. "We'd map cloud-specific controls directly to existing policies and create consistent patterns with automated verification."

He showed examples from their discovery work: IAM policies with least privilege, automated security scanning, and standardized logging configurations.

The discussion continued, with Isabel's team probing the details of the proposal. Throughout, Gary emphasized that the committee wouldn't replace existing controls but provide the structure to translate them effectively to cloud environments.

As the meeting concluded, Isabel seemed cautiously supportive. "This could work with executive backing and verifiable outputs. I'll need to see committee charters, control mappings, and regular compliance reporting."

"Of course," Gary agreed.

"Given the regulatory implications," Isabel added, "I'd expect quarterly updates to the Audit Committee."

"That's reasonable," Charles affirmed. "This is a priority for the entire organization."

As they left, Cathy pulled Gary aside. "That went better than expected."

"Isabel's concerns are valid," Gary replied. "We need more structure around our cloud controls. A committee gives us that while maintaining the collaborative approach that's working."

Day 167: The Cloud Excellence Committee

"So we're really doing this," Sarah said, studying the proposal Gary had just finished refining on his laptop.

A week after the meeting with Isabel from Audit, Gary and Sarah were putting the final touches on the Cloud Excellence Committee charter. What had started as a response to compliance concerns had evolved into something more ambitious as they incorporated feedback from across the organization.

"It's not what I originally envisioned," Gary admitted, "but it might be even better."

On screen was the formal charter for the Cloud Excellence Committee (CEC). A Shared Services team from central IT that would bring together Security, Operations, Finance, Asset Management, Identity Management, and Networking to develop consistent cloud standards and practices. Unlike previous approaches, where each central IT function created requirements independently, this committee would work collaboratively to align their guidance.

"Do you think the central teams will actually collaborate?" Sarah asked. "Everyone's been pretty protective of their domains."

"That's the big question," Gary replied. "Isabel's audit findings give us the executive mandate, but for this to work, we need commitment of time to focus on cloud across Shared Services in a unified way, not just as a side project."

The remediation efforts had continued over the past week, with some progress on priority issues. Sarah and Adam had been deeply involved, but it was noticeably pulling them away from their regular responsibilities. Getting other teams involved consistently, or at all, had proven challenging.

"This committee could help bring cloud as a first-class citizen for Shared Services," Gary continued. "Start prioritizing and putting actual capacity behind it, instead of just my small team trying to coordinate everything."

"I just got word that Charles wants us to co-present this at tomorrow's Executive Technology Committee meeting," Gary said. "He's been having one-on-one conversations with each department head over the past few days to get their input on the charter."

"Smart move," Sarah said. "No surprises in the meeting means no public objections."

"Just cautiously optimistic. We've done the groundwork. The collaborative approach has proven itself with several teams. The audit requirements give us regulatory justification. And we're proposing an enabling framework, not a controlling one."

Gary's phone buzzed with a message from Cathy, confirming her support for the proposal. That was significant. Security had traditionally maintained its independence from other central functions, developing requirements that weren't always aligned with Operations or Finance.

"Cathy is on board," Gary told Sarah. "That's a good sign."

An hour later, Gary found himself meeting with Oscar, who had questions about how the committee would work with existing operational processes.

"So this committee doesn't replace our existing teams?" Oscar asked, reviewing the charter.

"No, it coordinates how Shared Services approach cloud standards and practices," Gary explained. "Right now, Security has one set of requirements, Operations has another, Finance has cost guidelines, and Asset Management has inventory procedures. They're not always consistent."

Oscar nodded in understanding. "So instead of each central team publishing their own requirements independently, we'd work together to create unified guidance?"

"Exactly," Gary confirmed. "When divisions get different, sometimes conflicting, requirements from different central teams, they end up ignoring all of them. We need to present a unified approach."

"This would also help with our capacity issues," Gary added. "Instead of each team trying to solve cloud challenges separately with limited resources, we can pool our expertise and focus on the highest impact initiatives."

"What about Networking and Identity Management?" Oscar asked. "They haven't been as involved in our recent work, but they're critical for cloud operations."

"They're absolutely part of this," Gary agreed. "As we move beyond immediate security fixes into more strategic improvements, we need those teams more than ever. Identity for connecting cloud access to our corporate identity systems, Networking for extending our on-premises connections to cloud environments."

"Makes sense," Oscar acknowledged, offering his support for the charter.

The next day, Charles opened the Executive Technology Committee meeting by introducing the CEC proposal. Unlike the divisional town hall meetings, this session focused specifically on the central IT leadership who needed to align their Shared Services teams. The room included Cathy and the heads of each central IT function.

"Before Gary walks us through the details, I want to emphasize that this initiative has board-level visibility following Isabel's audit," Charles began. "I've spoken with each of you individually about how this committee will help us coordinate our cloud efforts more effectively while addressing our resource constraints."

He nodded to Gary to continue with the presentation.

"The account inventory and remediation work has made significant progress addressing immediate security concerns," Gary began, displaying metrics on identified vulnerabilities and completed controls. "But as Isabel's audit highlighted, we need a more sustainable approach from Shared Services."

He outlined the proposed committee structure, emphasizing how it would bring central functions together to develop consistent approaches rather than each team operating independently.

"This isn't about creating new requirements," Gary emphasized. "It's about aligning existing ones across Shared Services and ensuring they work together coherently. Right now, divisions receive different, sometimes contradictory, guidance from Security, Operations, Finance, and other central teams. The Cloud Excellence Committee will coordinate these efforts."

Gary highlighted how the committee would help address capacity constraints. "Rather than each department trying to tackle cloud with part-time resources while juggling other priorities, this committee gives us a mechanism to focus our collective expertise on the most critical needs."

The questions focused primarily on how this committee would interact with existing central teams and operational processes. Unlike previous meetings with division leaders, this discussion centered on how Shared Services would collaborate internally.

Charles closed the discussion with firm support. "This committee addresses a critical gap in how central IT approaches cloud operations. It has my full backing, and I expect all central functions to participate meaningfully." He looked around the room at the Shared Services leaders. "This isn't optional. Isabel's audit has made cloud controls a board-level concern. The Cloud Excellence Committee gives us a practical way to address it while making life easier for the divisions."

The charter was approved unanimously, with a directive to begin regular meetings the following week.

As Gary left the meeting, he felt a sense of accomplishment mixed with the weight of new responsibility. They had secured formal approval and executive backing, but now they had to deliver.

"Congratulations," Cathy said, catching up with Gary in the hallway. "Getting unanimous approval from the Shared Services leaders is no small feat."

"Thanks," Gary replied. "Amazing how quickly things happen when the work gets done before the meeting." He smiled slightly. "Now comes the hard part, making it actually work."

As Gary was driving home, he reflected on how their approach had evolved. They had moved from crisis response to structured improvement, and now to formal coordination across central functions. The Cloud Excellence Committee wasn't just another bureaucratic layer. It was a framework for collaboration that could transform how Goliath's Shared Services approached cloud operations.

Day 174: Executive Sponsorship

"Check your email," Sarah said, poking her head into Gary's office. "Charles just sent something to all of IT."

Gary opened his inbox to find a message from the Head of Technology Operations at the top of his unread list. The subject line was simple but impactful: "Cloud Excellence Committee (CEC) - Executive Announcement."

As he read through the message, Gary felt a growing sense of surprise. This wasn't the usual corporate announcement filled with vague platitudes and general encouragement. It was detailed, specific, and most importantly, it put Charles's full executive weight behind the cloud transformation efforts.

"The Cloud Excellence Committee represents a critical evolution in how we approach cloud at Goliath Federal," Charles had written. "This isn't just another initiative, it's a fundamental shift in how we operate."

The message went on to outline the committee's charter, its relationship to the ongoing remediation work, and the expectation that all central IT functions would actively support its efforts. Charles had specifically highlighted the audit findings as evidence of why better coordination was essential.

Most surprisingly, Charles had included a direct commitment of his own time and attention: "I will be personally reviewing the committee's progress on a bi-weekly basis and reporting outcomes to the Executive Leadership Team monthly. This has my highest priority, and I expect it to have yours as well."

"Wow," Sarah said, reading over Gary's shoulder. "I've never seen Charles put himself on the line like this for an IT initiative."

"Me neither," Gary replied, still processing the implications. The announcement transformed the Cloud Excellence Committee from just another IT project to a visible, executive-sponsored priority.

His phone buzzed with a meeting invitation: "CEC - Executive Briefing" with Charles, scheduled for that afternoon. The note simply read, "Let's align on next steps."

When Gary arrived at Charles's office later that day, he found him surrounded by printouts of cloud architecture diagrams and security reports.

"Come in, Gary," Charles said, gesturing to a chair. "I've been reviewing the documentation from the cross-divisional work and the committee charter. I want to make sure we're aligned on where this is headed."

Gary had worked with Charles for months now, and while Charles had been supportive in meetings and provided air cover during crises and audit discussions, this level of direct and public engagement was unprecedented. After all, Charles had thrown Gary out there alone for the town hall announcement months ago without providing any executive cover.

"Your announcement this morning was more detailed than I expected," Gary ventured. "It sent a strong message across the organization."

Charles nodded. "That was intentional. I've been providing guidance and support behind the scenes, but now it's time to make this a top-down priority. The work has proven its value, and formal executive sponsorship will help drive adoption across the organization."

He gestured to the security reports on his desk. "The audit findings were a wake-up call at the board level. They're asking hard questions about our cloud controls, and I need to show decisive action."

Charles continued, "But more importantly, I've been impressed with how much progress and momentum the team has made working with the divisions and overcoming challenges. We're better positioned than ever to form an official central cloud service to support the divisions. It's really the next phase to formalizing the importance of our cloud organization."

"The challenge now," Charles continued, "is institutionalizing that approach. The CEC is a good start, but it needs executive backing to be effective. Central teams respect my authority, but they also know how initiatives come and go. I needed to signal that this one is different."

As the meeting wrapped up, Charles had one more surprise. "I'm going to be creating dedicated positions to support this effort," he said. "Until now, everyone's been balancing this work with their day jobs, which isn't sustainable. I'll be working with each department head to formalize cloud roles and allocate proper resources."

"That would be a significant help," Gary replied. "The team has been stretched thin trying to juggle their regular responsibilities with this initiative."

"And I'm elevating the CEC's status," Charles added. "It will now report directly to me as a top organizational priority, with you as the program leader coordinating across departments."

"That's a big change," Gary managed, caught off guard.

"It reflects the importance of this work," Charles explained. "And frankly, it removes some of the bureaucratic barriers that have slowed progress. You've demonstrated the ability to drive cross-functional collaboration. I want to give you the organizational backing to match that responsibility, while everyone maintains their departmental roles."

After the meeting, Gary sat quietly, processing what had just happened. Charles wasn't just supporting the cloud initiatives anymore. He was actively elevating and formalizing them. This shift would have significant implications, providing much-needed authority and visibility while also raising expectations considerably.

Day 180: Unexpected Ally

Gary was reviewing the first draft of account consolidation recommendations when his phone buzzed with an unexpected message from Dennis: "Got time to talk about cloud organization?"

After reading it twice to make sure he wasn't imagining things, Gary replied with a meeting time. Dennis had been cordial since the security incident and had participated in the cross-divisional work, but he certainly hadn't sought Gary out for additional initiatives.

An hour later, Dennis arrived at Gary's office, laptop in hand and expression serious.

"Thanks for making time," Dennis said, settling into a chair. "I've got a situation I think you might be able to help with."

Gary waited, curious but cautious.

"We just got an automated alert from AWS about unpaid bills on three accounts we didn't even know existed," Dennis explained. "Turns out they belonged to a developer who left six months ago. His corporate credit card expired, and now AWS is threatening to suspend the accounts."

"Any idea what's running in those accounts?" Gary asked.

Dennis shook his head. "That's the problem. We're not even sure if they're being used for anything important. And this isn't the first time this has happened." He pulled up an email on his laptop. "Last month, we found two other accounts the same way. It feels like we're playing whack-a-mole."

Gary nodded, recognizing the pattern they'd seen elsewhere. "How are you handling it?"

"Right now? Scrambling to figure out what's in them before they get shut down," Dennis admitted. "But honestly, between this and all the remediation work, I'm spending more time on account management issues than on actual development."

Dennis shifted in his chair. "I was talking with Barry last week about our billing challenges, and he mentioned the CEC is working on account organization solutions. Something about consolidating under a master organization for billing and management. I'm wondering if that approach might help us."

This was an opening Gary hadn't expected. Dennis's team had been one of the first to establish shadow IT operations precisely to avoid central oversight.

"It definitely could," Gary confirmed. "Consolidating under AWS Organizations would give you centralized billing and basic account management."

"So no more surprise accounts appearing when credit cards expire," Dennis said.

"Exactly. We'd have visibility as soon as they're created, rather than discovering them months later."

Dennis nodded slowly. "I'd like my team to pilot this approach."

Gary blinked twice and smiled. "Wait, did I hear you correctly? You're volunteering?"

"Don't look so shocked!" Dennis said with a laugh. "This is becoming a real operational headache for us, and I'd rather be proactive than wait for the next surprise account to surface. Besides, as one of the first teams to dive into AWS, we probably have the most... let's call it 'distributed' account structure."

"That's a diplomatic way of putting it," Gary said with a chuckle. "But are you sure? This would mean central visibility into your cloud accounts."

"Look," Dennis said frankly, "I still believe in team autonomy for development. But basic account management? That's not where we add value. If anything, it's becoming a distraction from our real work."

He gestured to his laptop. "I'd rather have a solid foundation that lets us focus on building great applications than spend my time tracking down orphaned accounts and managing multiple bills."

Gary nodded, appreciating Dennis's pragmatism. "We'd be happy to have you pilot this. It's something the committee has identified as a foundational need anyway, so your team's experience would be valuable input."

"Good. And to be sure, we're just talking about account organization and billing consolidation, right? Not a wholesale change to how we build and deploy?"

"That's right," Gary confirmed, careful not to push too far too fast. "This is about the organizational framework, not your application development practices. Baby steps."

Dennis seemed relieved. "Perfect. I know there are other areas we'll need to address eventually, such as cost optimization, tagging standards, and all the other items on the remediation roadmap. But let's start here and see how it goes."

"Agreed," Gary said. "I'll have Adam reach out today to get the process started."

As Dennis prepared to leave, he paused. "You know, a year ago I would have fought tooth and nail against anything like this. Funny how things change."

"What changed for you?" Gary asked, genuinely curious.

Dennis thought for a moment. "My developers are wasting time managing cloud resources instead of building app features. Foundational stuff like paying for the cloud account, making sure identities are cleaned up when folks leave the company, wiring up the networking configurations, all can be a real operational headache."

He gestured to Gary. "If your team can handle that while being reasonably fast, that would be worth switching things up for. We want to build applications, not manage cloud accounts."

Day 187: Small Victories

"So where do we stand?" Charles asked, leaning forward in his chair.

A week had passed since the Cloud Excellence Committee's second meeting, and Charles had requested a status update from Gary and the team leads. The conference room held an unusual mix of participants: Gary, Sarah, Adam, Fran, and, most surprisingly, Dennis joining virtually from a business trip.

"We're making progress on multiple fronts," Gary began, displaying a simple dashboard showing their initial initiatives: account consolidation, security remediation, and tagging standards. "But we need to be strategic about our focus."

"Let's start with the account consolidation pilot with Dennis's team," Gary continued. "Adam, can you give us an update?"

Adam nodded. "We've successfully brought six of Dennis's accounts into the AWS master organization structure. The process was smoother than anticipated; it's primarily a billing relationship change without modifying account configurations."

"Any issues so far?" Charles asked.

"Nothing significant," Adam replied. "There was some initial confusion about how the consolidated billing would affect existing processes, but Finance is working through that."

Dennis unmuted himself on the call. "From our perspective, it's going well. The consolidated billing alone has already simplified our monthly reconciliation. And last week, we discovered another orphaned account through AWS's account team that we didn't even know existed."

Charles nodded approvingly. "Any interest from other divisions?"

"Jennifer from Retail Banking reached out yesterday," Gary reported. "She heard about Dennis's pilot and wants to discuss implementing the same approach for their accounts."

"Good," Charles said. "Organic adoption is exactly what we want to see. What about the other initiatives?"

Gary summarized their progress across security remediation and tagging standards, but noted a key challenge. "While we're making headway on multiple fronts, our resources are spread thin. Team members are juggling these efforts with their regular responsibilities, which is slowing our momentum."

"I see the same issue with the security work," Sarah added. "We've addressed 43 high-priority issues, but progress has stalled as team members get pulled back to their daily responsibilities."

"And the tagging standards are challenging to implement consistently," Adam noted. "Dennis's team has joined Peter's in piloting this approach, but retrofitting existing resources is a significant effort."

"We're starting with just tagging new resources," Dennis confirmed. "Tagging everything we already have would take several dedicated sprints, which is hard to prioritize against feature work."

Gary shared his observation. "We're spreading our limited resources across too many initiatives simultaneously. Some committee members are becoming bottlenecks as they try to support everything at once."

"What do you suggest?" Charles asked.

"While maintaining the broader control development work, we should focus our implementation efforts on one major initiative at a time," Gary explained. "The account consolidation shows the most promise. It addresses a fundamental pain point and delivers immediate value."

"While the technical implementation is relatively simple, inviting accounts into a cloud organization, the process changes around billing, and financial processes need time to adjust," Fran added. "Getting momentum with divisions and demonstrating success will build trust for future initiatives."

"So focus on account consolidation while keeping the other efforts at a pilot scale?" Charles clarified.

"Exactly," Gary confirmed. "It gives us visible wins while preventing burnout of our key resources. Once we've established a pattern of success with account consolidation, we'll have more credibility for the next priority."

Charles nodded thoughtfully. "Makes sense. Progress is about sustainability, not just speed." He glanced at his notes. "I'm encouraged by the interest from other divisions. Let's prioritize the account consolidation rollout while maintaining progress on the other fronts."

After the others had left, Charles asked Gary to stay behind.

"Now that we're a few weeks in," Charles said after the others had left, "what's your honest assessment on how the CEC is operating?"

"Better than I expected," Gary replied. "Having the team more focused on this work has made a real difference."

"And the challenges?"

"Even with more of their capacity, it's becoming more evident that we have a cloud skillset gap," Gary explained. "Most of the team is learning cloud as they go. They're doing great work, but I'm eager to get those cloud architect positions filled."

"What's the latest on the hiring process?" Charles asked.

"Two promising candidates are in final interviews," Gary replied. "They bring different strengths. One has a strong networking background, another has more DevOps experience. Finding good people with the right enterprise cloud experience takes time."

"Good to hear we're getting close," Charles said. "In the meantime, keep focusing on tangible wins with account consolidation. Nothing builds momentum like demonstrated success."

Day 194: Fran's Finances

She'd been working for a year now as Goliath Federal's Cloud Financial Analyst, and Fran still couldn't believe what she was seeing.

The spreadsheet on her screen showed cost data from Dennis's accounts that had joined their consolidated account structure. For the first time, she had actual visibility across multiple business units. Not estimates or projections. Real, accurate, comprehensive numbers.

When Fran took this job after eight years in traditional IT financial management, she thought she understood enterprise technology spending. You forecasted hardware purchases, tracked depreciation schedules, managed maintenance contracts. It was predictable, cyclical, and controllable.

Cloud changed everything.

Her first month at Goliath, she asked for the cloud spending reports and received seventeen different exports from various accounts, all in different formats, with inconsistent tagging. It was a financial analyst's nightmare.

She spent weeks trying to cobble together a coherent picture, only to discover she was seeing just a fraction of the actual spending. The shadow accounts, the forgotten test environments, the credit-card-funded projects, they were all invisible to her.

Now, with Dennis's accounts consolidated under a master organization, she could finally see the complete picture of their cloud spending. The patterns showed that 32% of their costs came from compute resources, but almost half of those were significantly underutilized. They were paying for premium storage when standard would meet their requirements, and their testing environment costs spiked at predictable intervals that aligned with their release schedule.

This was actionable intelligence, not just tracking spending, but identifying specific optimizations that could reduce costs without impacting performance.

Fran had prepared a report for tomorrow's meeting with Dennis, highlighting five opportunities that could save his division nearly $20,000 monthly. Not by cutting corners, but by using the cloud more efficiently.

That was the fundamental shift in financial management that she'd been trying to explain to her colleagues. In the traditional model, managing costs meant limiting spending. In the cloud model, it meant optimizing spending, using the right resources, in the right way, at the right time.

The coffee machine in Finance was broken again. As Fran walked to the one near the IT Operations team, she thought about how different her role had become since the Cloud Excellence Committee formed. Before, she was primarily a reporter of numbers. Now, she was increasingly an advisor, helping teams understand and optimize their cloud spending.

"Fancy seeing you over here," Oscar said as she reached for a mug. "Is the coffee on your side not good enough for you?"

"Broken machine," Fran explained. "But your coffee is better anyway."

"Just don't tell your Finance colleagues our secret," he stage-whispered. "We splurge on the good beans."

As they chatted, Oscar mentioned the account consolidation work. "Adam says we've got twelve divisions interested now. That's going to mean a lot more data coming your way."

"I know," Fran replied. "For the first time, we have a chance to do cloud financial management properly, moving from reactive to proactive. Instead of just tracking what we've spent, we can identify patterns, forecast needs, and recommend optimizations."

Oscar seemed intrigued. "Like what?"

"Like the Reserved Instance analysis I'm preparing for Dennis's team. By analyzing their usage patterns, I can show them how shifting from on-demand to reserved instances for their stable workloads would save 40% on those resources."

"And you couldn't do that before?"

"Not effectively. Without complete data, recommendations were always questionable. Teams would push back, saying 'you're not seeing the whole picture.' And they were right."

Back at her desk, Fran continued working on the Reserved Instance analysis. The challenge wasn't just identifying the savings, it was packaging the information in a way that made sense to technical teams who thought in terms of infrastructure, not finance.

Her phone buzzed with a message from Gary: "CEC tomorrow, 10 AM. Can you present initial findings from the consolidated accounts?"

The committee meetings had become a highlight of her week, a place where cross-functional collaboration actually happened rather than just being talked about.

Last week, when she shared the cost breakdown of the mobile banking app's cloud bill, the Security team spotted a suspicious spike in API requests. After investigation, they discovered a misconfigured security agent making unnecessary calls, a security issue that might otherwise have gone unnoticed.

The challenge now was scale. The account consolidation was giving them unprecedented visibility, but that visibility revealed just how complex their cloud environment really was. With hundreds of accounts across multiple providers and constantly changing usage patterns, comprehensive optimization seemed almost impossible.

That's why Fran had been researching specialized cloud financial management tools. She'd prepared a proposal for Gary outlining how they could help them analyze spending patterns and forecast costs more efficiently than their current manual processes.

But she was careful not to get ahead of herself. She'd seen too many initiatives at Goliath fail because they tried to do too much too quickly. The account consolidation was working precisely because it was delivering immediate value while building toward longer-term goals. And if they could centralize account management, they could get the appropriate access needed upfront for better reporting.

Dennis's team would be the test case. If she could demonstrate significant savings from the Reserved Instance strategy, it would provide a powerful example for other divisions.

Day 201: The Account Gateway Project

"This is a natural next step from the account organization work," Gary explained, sketching a simple diagram on the whiteboard.

Two weeks after prioritizing moving cloud accounts under a master organization, the CEC had turned its attention to planning the next strategic initiative. The account migration had gained enough momentum that they could begin thinking about what came next.

"First question, what do we call these things?" Gary asked with a slight smile. "AWS calls them accounts, Azure calls them subscriptions, Google calls them projects. Let's just say 'cloud accounts' for sanity's sake."

This drew a few chuckles around the room. These small inconsistencies between cloud providers were a constant source of minor frustration.

"Now that we're bringing accounts under a master organization, we need a standardized way to provision new ones," Gary continued. "Otherwise, we'll keep chasing orphaned accounts forever."

Adam had labeled it "The Account Gateway Project," and the name had stuck.

"What we're proposing is simple," Adam explained, taking over at the whiteboard. "A standardized intake process for new cloud accounts that captures essential information upfront that's hard to collect later like owner, purpose, environment type, and cost center."

Sarah nodded. "From Security, we'd ensure each new account has our monitoring tools deployed automatically. No more discovering accounts only after incidents occur."

Fran added, "Plus, all accounts would automatically join the consolidated billing structure, giving us immediate visibility into spending patterns."

The door opened as Peter arrived with Mike from Dennis's team. Mike had been invited to provide the developer perspective.

"So this is about standardizing how new accounts are created?" Peter asked, studying the whiteboard.

"Exactly," Gary confirmed. "Right now, it's too easy for teams to create accounts without proper configuration. We want to make the official process easier than using a credit card."

"The goal is to maintain team agility while ensuring control," Gary emphasized. "Application teams would still have full admin access to their accounts. We're just ensuring they start with the right foundation."

Mike raised a practical question. "What about the baseline configuration? How extensive would that be?"

"For the initial version, we're keeping it minimal," Gary answered. "Basic account settings, security visibility tools, and billing integration. We're not dictating networking configurations or internal architecture."

"That sounds reasonable," Mike said. "Teams still build environments how they need them, but central functions get the visibility they require."

The group identified key requirements: a simple request process, same-day provisioning, automated baseline configuration, and proper documentation of ownership.

"My team is..." both Peter and Mike started simultaneously, then stopped.

Peter smiled. "My team is setting up new development environments next month."

"And we're planning a new test environment for the mobile app," Mike added.

"We could both use this process if it's ready," Peter offered.

Gary tried not to look too pleased. Both Dennis's team and Peter's volunteering simultaneously was exactly the momentum he'd hoped for.

"We should have a basic version ready," Adam confirmed. "The core functionality will work, even without all the bells and whistles."

By the end of the meeting, they had outlined a simple approach: focus on the intake form and automation for Peter and Mike's upcoming AWS accounts, then scale to other teams and cloud providers.

As Peter and Mike left, Sarah stayed behind. "This addresses one of my biggest headaches," she admitted. "Finding accounts only after they're running sensitive workloads."

"It's a foundation," Gary agreed. "Not particularly exciting, but essential for everything that follows."

Day 208: Champions Assemble!

"Tell me about this Cloud Champions program," Charles asked.

Gary had just finished updating Charles on the Account Gateway progress and was about to leave when he raised the question.

"It's not a formal program yet," Gary admitted. "Just an idea we developed in yesterday's CEC meeting. While the committee is helping align our Shared Services teams, we've noticed we need better connections to the divisions themselves. The initial surge of engagement after the Jason incident has naturally tapered off."

"What we're seeing," Gary continued, "is that when changes come from central IT alone, there's often resistance or misunderstanding. But when someone from within the division that teams already know and trust advocates for those same changes, adoption happens naturally. These champions would be our bridge into the divisions, translating our guidance into terms that make sense for their specific teams."

"It's really about effective change management," Gary explained. "Having respected technical leaders in each division who understand both the central framework and their local context is proving essential. They can advise us on what will work in their environments while helping their teams understand the value of what we're building."

"That's the natural cycle of enterprise initiatives," Charles noted. "They start hot, then cool as the crisis fades."

"Exactly. But we've also noticed something interesting: in each division that's stayed engaged, there's usually one person who emerges as the internal advocate. They understand what we're trying to accomplish and become the go-to resource when questions arise."

"Like Peter and Mike," Charles observed.

"Exactly. They've been instrumental in getting their teams on board with the account organization and now the Account Gateway. But it's not just them. These people aren't assigned the role; they naturally step into it because they see the value."

The CEC had discussed how these individuals were critical to maintaining momentum as teams naturally drifted back to their regular responsibilities. The audit pressure wasn't as intense now that they had the CEC charter, but the work was far from complete.

"The idea is to recognize these individuals formally and give them tools to be more effective," Gary suggested. "Instead of central IT pushing everything, we'd build a network of these champions across the organization, people who already have credibility with their teams."

"I like this approach," Charles said. "It acknowledges that transformation happens through people, not just processes or technology. The timing works well too. The Technology Showcase next month might be a good opportunity to acknowledge these champions publicly."

The next day, the CEC held a special meeting with several early champions, including Peter, Dennis, Mike, and Jennifer.

"We've been thinking about recognizing what you all are already doing," Gary began. "After setting up that initial #cloud-champions channel in Teams, I've watched how you've naturally become the bridges between your teams and our cloud initiatives. What started as a place to ask questions has evolved into a real knowledge-sharing network."

"You know, I found myself chatting about the gateway patterns in the cafeteria just yesterday with a few team members," Peter revealed. "And Dennis and Jennifer were sitting next to us and ended up sharing what they were working on."

"That's exactly what we're hoping to encourage," Sarah replied. "You're already doing the work of champions. We just want to provide structure and recognition."

The champions shared their experiences and suggestions, helping shape the initial concept into something more concrete.

"What would help us most," Mike suggested, "is early access to new initiatives for feedback. It's easier to get our teams on board when we've had input from the beginning."

Jennifer nodded. "And some way to share solutions between divisions. Commercial Banking solved a cross-account access issue that I know Retail was struggling with, but there was no channel to share that knowledge."

By the end of the meeting, they had agreed to support the Cloud Champions network that had emerged in the Teams channel, focusing on connection rather than obligation. Rather than creating a formal program, they would simply provide these natural advocates with earlier access to initiatives, recognition from management, and resources to amplify their impact.

"This feels like a natural evolution of what's already happening," Adam observed as they finalized the proposal. "We're just making it official and providing some structure."

The meeting concluded with a sense of progress. What had started as a simple informal communication channel had created unexpected connections throughout the organization. The Cloud Champions weren't just linking central IT with the divisions, they were connecting the divisions with each other. Teams facing similar challenges were sharing solutions in ways they never had before, proving that the most effective communities form organically around shared challenges.

Day 215: First Light

"It's not much to look at yet," Gary cautioned as he pulled up the new cloud report on the projection screen. "And I want to emphasize this is very much a work in progress."

The CEC had gathered for their regular meeting, but today featured something new: the first version of what Adam was calling their "Single Pane of Glass" for cloud visibility.

"We've been collecting data for weeks," Adam explained, taking over the presentation. "From the account organization work, security assessments, and cost analysis. Until now, all that information lived in separate tools and spreadsheets. This report brings it together in one place, at least what we have so far."

The screen showed a simple Excel layout: a basic inventory of known cloud accounts across AWS, Azure, and GCP, with color-coding by division. Some accounts displayed basic metadata like owner, purpose, and environment, while others showed question marks or "Unknown" for several fields.

"As you can see, we're still missing a lot of information," Adam admitted, pointing to the gaps. "We're working to collect metadata for existing accounts, but it's a manual process. The good news is that all new accounts coming through the Account Gateway process will have this information from day one."

Sarah examined the limited security metrics with interest. "Even with the gaps, I can start to see which accounts have the most critical findings. That's helpful for prioritization."

"How much of this is automated?" Oscar asked, noticing some of the timestamps were several days old.

"Very little right now," Adam replied honestly. "It's mostly manual exports from different tools that we're combining weekly. We're still figuring out how to connect systems that weren't designed to talk to each other."

Gary nodded. "We wanted to get something visible to demonstrate the concept, even if it's still evolving."

Despite its limitations, the report was already revealing patterns that had been invisible when the data lived in separate silos.

"The value isn't just in having this information centralized," Gary pointed out. "It's about what teams can do with it. For example, look at these security posture indicators. Even with incomplete data, teams can see how they compare to their peers."

Fran nodded in agreement. "It's the same with cost metrics. Having this comparison creates a natural incentive to improve spending patterns."

Gary clicked through to a detailed view for one account. "Right now, we have basic inventory and metadata where available. We're manually compiling this weekly while we build better connections between our systems."

Oscar studied the operational metrics with interest. "I believe the engineers will appreciate a show back, a quantitative assessment of where they stand."

"That's exactly what we were thinking," Gary agreed. "Not to shame anyone, but to provide context for improvement. When a team can see the bigger picture beyond just their own accounts, it changes how they think about their approach."

Charles had been quiet until now, taking in the presentation. "How are we handling access to this report? Some teams might be uncomfortable with others seeing their data."

"The dashboard shows high-level metrics that aren't sensitive but haven't traditionally been shared across divisions," Gary explained. "More detailed reports are available in shared folders with appropriate access controls if teams need to dive deeper."

"And what about the division teams?" Charles pressed. "Do they have access?"

"That's what we wanted to discuss today," Gary replied. "We think division teams should see their own data in full detail, plus summary information about other divisions for context. It would help them understand the overall landscape and give the Cloud Champions specific areas to focus on."

This represented a significant shift in Goliath's traditional approach to information sharing. Historically, divisions rarely saw data from other parts of the organization.

"I support that approach," Charles said after a moment's consideration. "Breaking down these information barriers is essential if we want to drive consistent practices."

With Charles's endorsement, the conversation shifted to how they would roll out access to the report. They decided to start with the Cloud Champions for feedback before gradually expanding to broader technical teams.

"What do we need for the next iteration?" Charles asked as the discussion wrapped up.

Gary outlined their roadmap: "We'll continue improving the data quality while working toward more comprehensive insights. The quarterly Technology Showcase is in two weeks; that would be an ideal opportunity to demonstrate our progress."

"Perfect timing," Charles noted. "We can showcase this alongside recognizing the Cloud Champions."

After the meeting, Gary found himself reflecting on what they'd just accomplished. The dashboard itself was rudimentary, just Excel sheets with some formatting and basic charts. But its potential impact ran deeper. By making previously isolated information visible across teams, they were creating the foundation for natural behavior change. When teams could see how their practices compared to others, improvement became a matter of professional pride rather than compliance.

Day 222: Barry's Wall Cracked

Twenty years at Goliath, and Barry had seen them all come and go. New leaders. New initiatives. New buzzwords.

Barry had survived seven reorganizations, four strategic transformations, and more "game-changing" initiatives than he could count. Commercial Banking continued to deliver regardless, because they focused on their business, not the latest corporate fad.

But this cloud stuff... it was different.

Barry would admit, though never publicly, that Gary's team was actually making progress. Not the sweeping transformation they thought they were creating, but meaningful improvements in specific areas. The account organization had simplified their billing. The Account Gateway had streamlined provisioning. Even their basic dashboard was revealing insights they didn't have before.

Still, Barry remained skeptical. Twenty years of experience had taught him the value of careful, measured approaches.

"Barry, do you have a minute?" Dennis, his FinTech lead, stood in Barry's doorway, tablet in hand.

"Of course, come in." Dennis had proven his value over the years with his technical expertise and business focus.

He closed the door and sat down. "I've been reviewing the CEC's latest dashboard. There's something you should see."

He pulled up a chart showing security posture scores across divisions. Commercial Banking was in the bottom third, behind Retail Banking and even Wealth Management.

"This doesn't make sense," Barry said, frowning. "We have the most sophisticated security controls in the bank."

"That's what I thought too," Dennis agreed. "The scoring is heavily weighted toward implementing the standard security monitoring tools from central IT. Since we've maintained our own approach, we're scored lower despite having equivalent or better protection."

This was exactly what irritated Barry about centralization efforts. They inevitably created arbitrary standards that valued conformity over actual results.

"Has anyone else seen this?" Barry asked.

"It was just released to the Cloud Champions yesterday. Charles is planning to showcase it at the quarterly technology review next week."

Having Commercial Banking appear less secure than other divisions was unacceptable. They prided themselves on maintaining the highest standards.

"What options do we have?" Barry asked.

"Three main choices," Dennis replied. "We protest the scoring methodology, quickly implement their monitoring tools, or present our own parallel assessment."

Each had drawbacks. Fighting the methodology would make them look resistant. Rushing implementation wasn't practical. A parallel assessment might further isolate them.

"Let's take a different approach," Barry decided. "Schedule a meeting with Gary and Sarah. Let's understand their methodology and see if there's a way to accurately reflect our security posture without compromising our standards."

Dennis looked surprised. "You want to work with them on this?"

"Strategically, yes. If their dashboard is gaining traction, we need to ensure it reflects reality."

Barry explored the other dashboard metrics. In cost metrics, they were middle of the pack. In operational excellence, near the top. Deployment frequency, also high.

This had been Barry's internal conflict since Gary arrived. He believed in division autonomy. But some degree of standardization made sense, particularly as their cloud footprint grew more complex. And he couldn't deny that the CEC had been effective at redirecting audit pressure toward central IT and away from individual divisions like his.

The next morning, Barry arrived at the meeting prepared to defend their approach while finding a constructive path forward. He was surprised to find Gary already engaged in conversation with Dennis, comparing notes on AWS capabilities.

"Barry, thanks for suggesting this meeting," Gary said. "Dennis was just sharing some of your custom security controls. There's impressive work here."

"We've invested significantly in our security posture," Barry acknowledged. "Which is why I was concerned about how we're represented."

Over the next hour, they had one of the most productive technical conversations Barry had had in months. Sarah acknowledged that Commercial Banking had implemented several advanced controls before central Security had even considered them.

"The challenge is standardizing the assessment across divisions," Sarah explained. "We need a methodology that drives adoption of baseline standards."

"I understand the need for baselines," Barry conceded. "But the methodology should recognize equivalent controls, not just specific implementations."

Gary had been mostly listening, taking notes. Now he spoke up. "Take network architecture. We know breaking apart existing network designs is challenging and costly. So instead of forcing everyone to use a pre-defined template, we should assess whether teams are achieving core principles like proper segmentation and least-privilege access. If a team's existing network design effectively isolates critical systems and limits potential breach impacts, that should count as meeting the intent of the control, even if it doesn't match our exact blueprint."

It was a reasonable suggestion that acknowledged both the need for standards and the value of division-specific innovations.

As they wrapped up, Gary made a comment that stuck with Barry: "The dashboard isn't about forcing conformity. It's about visibility. Sometimes that means standardizing how we measure things, but it doesn't have to mean standardizing everything we do."

It was a nuanced position, more sophisticated than Barry had expected from someone relatively new to Goliath.

Barry wasn't ready to fully embrace the CEC's agenda. But he was beginning to see areas where standardization genuinely added value without undermining their autonomy. The revised security scoring meant Commercial Banking would accurately appear near the top in the quarterly review. The jury was still out on the CEC, but Barry was becoming less certain of its failure.

His decision to collaborate rather than resist yielded a better outcome for Commercial Banking. Perhaps that was the lesson after twenty years at Goliath: knowing when to resist and when to collaborate served their interests better than blanket opposition.

Day 236: System Shock

The alert came at 3:17 AM.

By 3:30 AM, Gary was on the incident bridge call from his home office, trying to make sense of multiple voices talking over each other in barely contained panic.

"Can someone give me the latest status?" Gary asked, cutting through the noise.

Oscar took the lead. "Multiple production services down across three divisions. Started with an authentication failure in Commercial Banking, then cascaded to payment processing and customer account access."

"Any recent changes?" Gary asked.

"Commercial Banking deployed an identity federation update yesterday, but it passed all tests," Dennis replied. "AWS is also reporting degraded network connectivity in us-east-1a, starting about 20 minutes before our issues began."

As teams worked to diagnose the problems, Gary was struck by how quickly the response came together. Six months ago, they would have struggled just to identify which services were affected and who owned them. Now, with the account organization work and shared dashboard, they had immediate visibility into the impacted systems and their dependencies.

By 6:00 AM, they had established a workaround by manually redirecting traffic, but the root cause remained unclear. Gary had made his way into the office by 7:00 AM to coordinate the response in person. Customer-facing services began recovering, though internal applications remained impacted.

"We're still seeing authentication failures for several internal systems," Dennis noted around 7:15 AM. "Our primary authentication service should have redirected to the backup instances, but that's not happening properly."

Mike from Dennis's team discovered the issue. "I see the problem now. Our load balancer is working, but all the backend authentication servers are in the same AWS availability zone datacenter, which is the one experiencing issues. There's nothing healthy for the load balancer to route to."

Oscar pulled up the system topology from their central dashboard. "That's right. The older authentication service was deployed before we had visibility across environments. The load balancer is configured correctly, but all the resources behind it are in the us-east-1a availability zone only."

"Our payment services are affected too," Jennifer added from the conference bridge, "but only the older ones. The new microservices we deployed through the Account Gateway are still running without issues."

This was significant. The newer services had been deployed across multiple AWS availability zones for redundancy across cloud data centers, following the baseline documentation guidance, and remained operational throughout the incident. Only the older services with single points of failure were down.

By 10:30 AM, with AWS services stabilizing, they had restored most critical functions. Full recovery of all internal systems took until early afternoon.

Charles joined the post-incident review at 4:00 PM. "What's the root cause analysis?" he asked.

"A high availability design that wasn't actually highly available," Oscar explained. "The older services were set up with load balancers for redundancy, but the backend resources were all deployed in the same availability zone. When that availability zone experienced issues, the entire service failed despite the redundant design."

"The services using the newer multi-availability zone patterns remained operational throughout the incident," Gary added. "Only the older implementations experienced extended downtime."

Dennis nodded. "Our authentication service was supposedly highly available, but its redundancy was all within a single physical location. We never considered that scenario in our testing."

"What made the difference in our response," Sarah noted, "was having immediate visibility into which systems were affected and understanding their interdependencies. Six months ago, we would have spent hours just figuring out what was happening."

Charles looked around the room. "Alright, let's capture the key actions. We need to review our existing services against these multi-availability zone patterns, and update our baseline documentation to make this guidance more prominent."

After the meeting, Gary pulled Nancy from Networking and Adam aside to discuss the multi-availability zone design lessons from the incident.

"We need to expand the baseline configuration," Nancy suggested. "Right now we're focusing on account settings and basic monitoring. We should include network architecture patterns and multi-availability zone deployment templates as standard components."

Gary nodded. "Specifically, what network elements are critical for teams to get right from the start?"

"The VPC configuration is key," Nancy explained. "Pre-defining IP ranges that align with our existing network, setting up subnets across multiple availability zones for resilience, and ensuring proper routing and interconnects. But we don't want to over-engineer."

Adam chimed in, "Exactly. We're providing the foundational network infrastructure that's complex and time-consuming for teams to set up correctly."

"Not application-specific components," Gary confirmed. "We're not defining load balancers or specific application architectures. Our goal is to give teams a solid, secure network foundation that makes it easy to deploy quickly and safely."

"And our controls can still verify things like multi-availability zone configurations for critical services," Nancy added, "without us pre-defining every architectural detail."

"A landing zone," Adam said, "with these best practices already implemented."

As Gary left the meeting, he reflected on how their approach was evolving. What started as a simple account provisioning process was growing into something more comprehensive, a foundation for resilient cloud architecture that teams could build upon with confidence.

Day 243: Blame Game

"I thought we'd moved past this," Gary muttered as he scanned the latest email thread.

A week after the major cloud outage, what had started as a constructive post-mortem process had devolved into finger-pointing. The collaborative spirit during the incident had evaporated once the immediate pressure was off.

The executive summary for the Executive Technology Committee had triggered the shift. Each division had carefully worded their contributions to cast themselves in the most favorable light.

Commercial Banking minimized the role their authentication service played in the cascade failure. Retail Banking highlighted that their newer services had experienced minimal disruption. Wealth Management pointed out that their services were the first to recover.

What was missing was the core insight: the services that recovered most efficiently shared certain architectural characteristics, regardless of which division owned them.

His phone buzzed with a message from Sarah: "Have you seen the revised incident report?"

"Just reading it now," Gary replied. "Not exactly the unified message we discussed in the technical debrief."

"Commercial Banking removed the entire section about their load balancers pointing to single availability zone resources," Sarah wrote.

The post-mortem revisionism wasn't surprising, but it was disappointing. During the outage, engineers from different divisions had worked together seamlessly. But once the crisis passed, the institutional reflexes reasserted themselves.

At the Executive Technology Committee meeting, Charles opened with his usual directness. "I want to address last week's cloud outage. Gary, can you give us a brief overview of what happened and what we've learned?"

Gary focused on the technical facts. "The root cause was a partial AWS regional degradation that affected our authentication services, which then cascaded to dependent systems. The event revealed specific patterns that worked better than others during recovery."

He displayed a simple diagram showing the sequence of failures and recovery times across different services.

"We identified three specific architectural patterns that made a significant difference," Gary explained. "First, services with resources distributed across multiple availability zones continued functioning even when one availability zone was impaired."

"Second, services with properly configured load balancers recovered much faster than those with single points of failure. And third, services with documented service relationships allowed teams to quickly identify which systems would be affected by the authentication failure."

Barry shifted in his seat. "That's one interpretation. Another is that the services that recovered fastest were simply those with the most experienced engineering teams managing them."

Jennifer spoke up next. "I'd like to point out that our newer services were among the first to recover. We didn't have years of legacy implementations to maintain."

"I appreciate everyone's perspectives," Charles intervened. "But I'm more interested in what we're going to do differently going forward. Gary, what are the key recommendations?"

Gary presented two action items: "First, getting our new landing zone concept released for guidance to emphasize multi-availability zone network architecture deployments. And second, enhancing our monitoring to detect when teams are only using single availability zone configurations that pose resilience risks."

Gary added. "For new accounts, we'll provide configuration templates that implement these patterns themselves. For existing accounts, we'll add dashboard indicators to highlight resilience risks."

After the meeting, Charles asked Gary, Sarah, and Oscar to stay behind.

"That was quite the diplomatic navigation," Charles observed.

"I tried to focus on the technical patterns rather than making it about whose approach was better," Gary explained.

"The technical teams who were responding to the outage were remarkably collaborative," Oscar noted. "It was only afterward that the positioning started."

"That's because during a crisis, everyone is focused on the same goal," Sarah added. "Once that immediate pressure is gone, old habits reassert themselves."

Charles considered this. "So the challenge is maintaining that sense of shared purpose beyond crisis moments."

"Exactly," Gary agreed. "And that's why these iterative improvements to our baseline are so important. Each incident teaches us something new that we can incorporate into our guidelines and tools."

Day 250: Unexplored Territory

Gary studied the organization chart on his screen, focusing on the divisions they hadn't yet engaged in their cloud initiatives. While they'd made significant progress with forward-leaning teams like Peter's and even gained traction with more traditional divisions like Commercial Banking, several large portions of Goliath remained untouched by their cloud work.

"Completely unexplored territory," Gary murmured.

"What's that?" Sarah asked, stopping by his office.

"I've been mapping our engagement across the organization," Gary explained, turning his screen toward her. "See these highlighted areas? Risk Analytics and Enterprise Data Services teams have been completely absent from all our cloud initiatives. No Cloud Champions, no Account Gateway participation, not even attendance at the Technology Showcase."

Sarah studied the diagram. "You're right. I hadn't noticed how concentrated our progress has been with certain divisions."

"It's a blind spot," Gary acknowledged. "We've been so focused on the cloud-forward teams and navigating resistance from traditional IT groups that we've completely missed these areas."

"But why would they be on our radar? If they're not using the cloud yet."

"That's just it," Gary interrupted. "According to the last AWS billing consolidation, there are at least a dozen accounts associated with RA email domains coming from the acquisition last year. They're doing something in the cloud, just completely disconnected from everything we're building."

"These aren't resistance cases like we've dealt with before," Gary continued. "These are teams that might not even know what we're doing. They may not have the cloud expertise of Peter's team or the established practices of Commercial Banking. They're operating in isolation, likely figuring things out as they go."

"So what's the approach?" Sarah asked. "Same as with the other divisions?"

Gary shook his head. "I don't think so. The Account Gateway and resilience patterns we've been developing assume a baseline of cloud understanding. These teams might need something more fundamental."

This was a different challenge than they'd faced before. With cloud-forward teams like Peter's, they'd focused on enhancing and standardizing already solid practices. With more traditional divisions like Commercial Banking, they'd navigated complex resistance rooted in established ways of working. But these unexplored divisions might be at the very beginning of their cloud journey, perhaps using cloud services with minimal expertise or strategic direction.

"I'm meeting with the RA leadership this afternoon," Gary said. "Charles helped arrange it after our discussion about engagement gaps. The good news is that we finally hired a new cloud architect who is starting next week, so the timing is perfect."

The meeting with RA was eye-opening, though not in the way Gary had expected. Instead of finding a team hesitant about cloud, he discovered they were actually enthusiastic but overwhelmed.

"We've been trying to enhance our fraud detection capabilities for over a year," explained Robert from Risk Analytics, their director. "Our current infrastructure simply can't handle the AI workloads we need to develop effective models, and we need access to the latest machine learning services."

"When did you start creating cloud accounts?" Gary asked.

"About a month ago," Robert acknowledged. "Several of our data scientists have set up environments for specific projects. But honestly, we're stumbling through it without a specific strategy."

As the conversation continued, Gary realized this wasn't resistance or disinterest. It was a fundamental gap in cloud understanding and expertise. The team knew they needed cloud capabilities but lacked the knowledge to implement them effectively.

"Why haven't you engaged with the Cloud Team before now?" Gary asked.

Robert looked slightly embarrassed. "Frankly, we assumed your focus was on the application teams and infrastructure migration. Our fraud detection platform has unique requirements that didn't seem to fit the standard approaches."

This perception, that cloud management meant standardization that wouldn't accommodate specialized needs, was a critical insight. It explained why some divisions might be steering clear of engagement rather than actively resisting it.

"What would help you most right now?" Gary asked.

"Honestly? Architecture support," Robert replied without hesitation. "My team needs guidance on implementing AI services in the cloud, securing access to sensitive data, integration with our on-premises systems, and best practices for model development and deployment."

Later, Gary reflected on the conversation with Robert. RA wasn't avoiding cloud management initiatives; they were struggling with cloud adoption itself and had exciting new business initiatives that were a perfect use case for the cloud.

Day 257: Charles's Strategic Pivot

The conference room was quiet as Charles reviewed his notes before the IT Steering Committee meeting. As Head of Technology Operations, these quarterly gatherings with division IT leaders and business stakeholders were his opportunity to shape Goliath's technology direction. Today, he was taking a calculated risk.

Gary's discovery of untapped divisions like Risk Analytics and Enterprise Data Services had given Charles an idea. For months, they'd positioned cloud as an infrastructure initiative, focusing on migration, standardization, and operational efficiency. But these data-intensive teams needed something different. They needed business capabilities, not just infrastructure.

"Let's get started," Charles began as the attendees settled in. Instead of opening with their standard metrics on data center utilization and security compliance, he started with a question.

"Robert, could you share the current challenges your RA team is facing with the fraud detection models?"

Robert looked surprised but quickly composed himself. "We've been working on enhancing our fraud detection capabilities for a number of months now. The pattern recognition capabilities we need require computational power our current infrastructure can't provide, particularly for the AI models we're trying to implement."

"And what's the impact of that limitation?" Charles pressed.

"We've had to simplify the models and extend timelines. What should be a three-month development cycle has stretched to nine months, and we're still not where we need to be to effectively combat emerging fraud patterns."

Charles nodded and turned to Eric from Enterprise Data Services. "And what about EDS? I understand you've been working on scaling your data analytics platform?"

"Similar situation," Eric confirmed. "We've been trying to modernize our data platform for over a year now. Our on-premises data warehouse simply can't scale to handle the volume and complexity of analysis our data scientists need. We need to leverage more high-performance computing while maintaining connections to our existing systems."

"Thank you both," Charles said, standing to take control of the presentation. "These challenges highlight something important about our cloud journey so far. We've been focused on infrastructure migration and standardization, but we haven't fully leveraged cloud capabilities to address specific business challenges like these."

Charles brought up a slide showing a reference architecture Gary's team had begun developing with RA, a hybrid platform that leveraged cloud AI services while maintaining secure connections to existing on-premises data sources.

"This is just one example. Instead of asking divisions to adapt to standard patterns, we're adapting our patterns to specific business needs, data analytics, risk modeling, and fraud detection, while maintaining the security and operational foundations we've established."

Walter from Wealth Management raised a question. "This sounds like consulting. How is it different from what divisions are already doing with their own resources?"

"Three key differences," Charles replied. "First, these specialized patterns are built on the foundations we've already established. Second, they're designed for reusability across divisions with similar needs. And third, we've been growing our cloud architecture team with deep expertise in these domains."

Robert looked interested but skeptical. "We've tried working with central IT on specialized needs before. It usually ends with us having to adapt to standardized approaches that don't meet our requirements."

"That's exactly what we're changing," Charles explained. "This isn't about forcing standards, it's about creating specialized patterns for your specific needs, with guidance on what has worked for others in the organization. We've learned a lot about the balance between standardization and flexibility."

By the end of the discussion, they had agreed to proceed with two initial pilots, one with Robert's RA team on their AI-powered fraud detection capabilities, and another with Eric's EDS team on their scalable analytics platform.

After the meeting, Gary joined Charles in his office to debrief.

"Two pilot engagements is a good start," Charles told him.

"It's great, starting with just a couple of projects gives us a chance to refine our approach before scaling it," Gary replied.

"The key is showing concrete business value," Charles emphasized. "Not just technical metrics or infrastructure efficiency, but actual business capabilities delivered faster than would have been possible otherwise."

"What I found most interesting," Charles reflected, "was how differently Robert and Eric approached this compared to our typical discussions with application teams. They're less concerned about control and more focused on capability. They just want to solve their business problems."

As Gary left, Charles considered their evolving role. They weren't becoming the experts with all the answers, but rather the connectors, bringing together insights from different divisions, identifying common challenges, and helping teams learn from each other. Peter's team had solved a networking challenge that could help Dennis. Dennis's cost optimization approach might benefit Jennifer's team. Their job wasn't to dictate solutions, but to create channels for knowledge sharing.

By focusing on business capabilities rather than infrastructure standards, they were meeting divisions where they were, not where they thought they should be.

Day 264: Turning Point

"So where do we go from here?" Gary asked, looking around at the Cloud Excellence Committee members gathered in the conference room.

Since launching their cloud dashboard, a pattern had emerged. New accounts provisioned through the Account Gateway showed significantly higher compliance rates than older environments. The baseline configurations implemented key controls from the start, while existing accounts struggled with manual remediation.

"Let me show you something telling," Adam said, displaying a comparison chart. "Accounts created through the Account Gateway process show 87% compliance with our baseline standards. Pre-Gateway accounts average only 42%, despite all our notification efforts."

"The dashboard is identifying issues effectively," Sarah explained, "but remediation rates for existing accounts remain concerning." Her chart showed most critical security alerts lingering unaddressed long after notification.

"The network architecture complexity is a major barrier," Nancy observed. "Most application teams lack the specialized knowledge to configure VPCs, security groups, and transit gateways correctly. They're developers, not network engineers."

"And with identity management," Isaac from Identity Management added, "teams are afraid to modify IAM roles and policies once things are working. They worry any change might break existing access patterns or applications, so they leave overly permissive configurations in place."

Fran nodded in agreement. "There's also alert fatigue. When teams see dozens of recommendations across security, cost, and operations, they don't know where to start or which to prioritize."

Gary studied the adoption patterns across their initiatives. The challenges were crystallizing: integrated cloud architecture was complex, teams lacked domain expertise in areas outside their core focus, they prioritized application development over infrastructure improvements, and they struggled to navigate the volume of recommendations.

"I think we've identified our critical gap," Gary said, moving to the whiteboard. "The Account Gateway provides minimum baselines and guidance, which works well for new accounts. But teams need more than alerts and documentation; they need expertise and support."

"What if we evolved our approach in two ways?" Gary proposed. "First, expand what's automatically configured in the Account Gateway, moving from basic landing zones to comprehensive safe landing zones. Second, provide more hands-on implementation support for existing accounts."

The energy in the room shifted as committee members connected this approach to their specific areas.

"Safe landing zones," Adam repeated thoughtfully. "I like that. Not just basic account setup but a comprehensive foundation that reduces the burden on application teams."

Isaac and Sarah exchanged glances. "We could significantly expand what we pre-configure centrally," Isaac suggested. "Beyond network architecture, we're talking about standard logging configurations, centralized monitoring setups, core IAM roles with least-privilege principles, and standardized security baselines that teams can build upon."

"Exactly," Nancy added. "We're creating a more robust starting point. The networking patterns we developed are just the beginning. We want to give teams a solid, secure foundation that handles the complex, foundational elements they shouldn't have to figure out themselves."

"And we'll continue providing specialized guidance for specific use cases," Gary noted, "like the public-facing service pattern we developed with Commercial Banking, and other patterns we have working on specifically with the Risk Analytics and Enterprise Data teams.

Charles, who had arrived partway through the discussion, nodded approvingly. "This aligns with what division leaders tell me. They're not resistant to standards, they're overwhelmed by them." He paused, considering his next words carefully. "Remember, in this organization, how you deliver is just as important as what you deliver. Focus on making the right way the easy way. If you want people to follow best practices, make those practices the path of least resistance."

"As the meeting concluded, Gary made a note to update Isabel from Internal Audit on their evolving approach. Her quarterly assessment of their cloud governance maturity was scheduled for next week, and these insights about implementation gaps would be relevant to her evaluation. The Audit Committee would want to know how they were addressing the remediation challenges she had highlighted in her previous report."

Day 271: Isabel's Assessment

"Trust, but verify." That was the unofficial motto of the Internal Audit team at Goliath Federal Bank, and it was one Isabel lived by. Unlike many of her peers who treated cloud environments like traditional data centers, she'd recognized that cloud required a different approach entirely. The dynamic, software-defined nature of cloud demanded a more pragmatic stance balancing control with innovation, compliance with momentum.

Isabel reviewed her findings in their GRC risk management platform one last time before heading to the quarterly Audit Committee meeting. She'd be presenting their assessment of the maturity of Goliath's cloud controls, and the stakes couldn't be higher after the authentication outage two months ago.

When Isabel first initiated this review, the situation was alarming. They had multiple divisions operating cloud environments with inconsistent controls, no central visibility, and significant compliance gaps. Their Internal Asset Protection Policies (IAPPs), for the most part, simply weren't translating to cloud environments in any coherent way.

What had changed since then was remarkable, though they still had significant work ahead. Her team had been working collaboratively with Gary's team rather than taking the traditional audit approach of rigid assessment after the fact. They'd prioritized establishing foundational elements first, such as visibility around inventory and configurations, setting up basic security and financial cost monitoring, and operational controls.

The report Isabel was presenting today showed genuine progress from "high risk" to "moderate risk" in their overall assessment. But more importantly, it showed the emergence of a sustainable approach to managing cloud environments.

It would be easy for Audit to simply declare the cloud "non-compliant" and shut things down until all required IAPP controls were documented and implemented. But that approach would hurt the bank's ability to innovate and deliver critical capabilities for their clients.

The elevator doors opened on the executive floor, and Isabel made her way to the boardroom where Elizabeth, the Audit Committee chair, welcomed her. As the remaining committee members arrived, she noticed Charles slip in and take a seat near the back.

Once everyone was seated, Isabel began her presentation.

"Last meeting, I presented findings that identified significant gaps in our cloud control environment," Isabel started, displaying a comparison of their previous and current risk assessments. "Today, I'll share how those gaps have been addressed and where risks remain."

"The most significant advancement has been in account visibility and organization," Isabel explained. "Through the Account Gateway and consolidation initiatives, we now have visibility into over 90% of our cloud accounts which have oversight from Shared Services, up from less than a third at the beginning of this assessment."

"Security controls show mixed progress," Isabel continued. "The identification of vulnerabilities has improved dramatically, but remediation rates remain inconsistent. Only 43% of critical findings are addressed within target timeframes."

"What's causing the delay in remediation?" asked Elizabeth.

"Implementation complexity is the primary factor," Isabel explained. "Teams understand what needs to be fixed but often lack the expertise to implement corrections properly. The CEC is addressing this by developing implementation tools and expanding their baseline configurations through what they're calling 'safe landing zones'. Which essentially is pre-configured cloud environments that meet core security and compliance requirements from the start."

After covering all assessment areas, the questions began.

"How does our approach compare to industry peers?" asked one director.

"We're behind the technology and financial services leaders but improving rapidly," Isabel replied honestly. "The focus on practical implementation tools rather than just policies is more advanced than many peers who are drowning in cloud security alerts they can't remediate."

"And what about regulatory expectations?" asked another. "Will this satisfy our examiners?"

"Our approach aligns with regulatory guidance," Isabel explained carefully, "but we're walking a tightrope. We're subject to many different regulatory bodies, each with their own interpretation of infrastructure controls. The Federal Reserve, Securities and Exchange Commission, etc., they will be scrutinizing our cloud migration with increasing intensity."

Isabel pulled up a detailed slide showing a heat map of regulatory findings from a recent assessment with on-premises infrastructure. "We have seventeen critical observations ranging from data protection, access controls, and change management in our environments. Historically, we would have months, sometimes over a year, to remediate these findings. But I am concerned that with the cloud's rapid evolution, regulators will demand faster, more comprehensive responses."

Elizabeth's expression grew more serious.

"We're taking a risk-based approach that prioritizes the most critical controls," Isabel continued. "The alternative would be to restrict cloud usage entirely until we have perfect controls, which would significantly impact our business capabilities. But make no mistake, we're under significant regulatory pressure. The controls you are implementing need to align with our IAPPs, not just follow industry best practices."

"What recommendations do you have for next steps?" Elizabeth asked.

"Three specific improvements," Isabel replied. "First, we need to demonstrate how our cloud environments satisfy the bank's IAPPs. Second, we need more defined implementation timelines and verification methods. And third, we need better remediation capabilities."

Isabel emphasized the importance of translating cloud controls to their established policies. "This is about connecting what we're doing in the cloud to what regulators expect to see. We need a practical approach that addresses security, financial, and operational risks while maintaining alignment with our existing control framework."

After the committee members dispersed, Charles approached Isabel in the hallway.

"Balanced assessment," he acknowledged. "I appreciate you recognizing the business impact of restricting cloud innovation."

"Audit's role isn't to block progress," Isabel replied. "It's to ensure we understand and manage risks appropriately. A risk-based approach that allows innovation while systematically improving controls serves the bank better than rigid compliance that blocks business capabilities."

As Isabel packed up her materials, she reflected on the journey ahead. The team's approach to implementing security controls first was promising, addressing their most immediate needs while building toward broader operational improvements. Their job wasn't to force the cloud into their existing framework, but to evolve their framework to effectively address cloud risks.

Day 275: Framework Emerges

"These audit findings are both a challenge and an opportunity," Gary said, reviewing Isabel's report with the Cloud Excellence Committee. Just a few days had passed since her presentation to the Audit Committee, and the CEC had gathered for a focused working session on their response.

"Isabel's assessment was fair," Sarah noted. "We've made significant progress on visibility and standards, but we're still struggling with consistent implementation and remediation."

Gary nodded in agreement. "The audit findings highlight that we need to connect our cloud work to the bank's formal control requirements."

"What exactly did the Audit Committee request?" Nancy asked.

"Three specific improvements," Gary explained, displaying Isabel's recommendations. "First, we need to demonstrate how our cloud environments satisfy the bank's IAPPs. Second, we need more defined implementation timelines and verification methods. And third, we need better remediation capabilities."

"The challenge with the IAPPs," Gary continued, "is that they were designed for traditional data centers. They specify what controls must be in place, but not how to implement them in cloud environments."

Charles, who had attended the Audit meeting, nodded. "The Audit Committee doesn't expect us to implement controls exactly as written in the IAPPs. They need us to show how different cloud implementations still satisfy the same control objectives."

"So we need a translation layer," Adam said. "Something that maps cloud-specific controls to the existing IAPP requirements."

"Exactly," Gary confirmed. "And there are a few approaches we could take."

"What about adopting industry standards like CIS or NIST?" Adam asked, referring to established cybersecurity frameworks. "Couldn't we map those to our IAPPs?"

"CIS Benchmarks are excellent for security controls," Gary acknowledged. The Center for Internet Security provided widely-adopted security guidelines that many organizations used as baselines. "They're prescriptive, focused on a manageable set of controls, and many of our existing security tools already support them."

Sarah nodded. "Our security monitoring platform already has CIS compliance built in. Using that as our security foundation would give us immediate visibility into our posture."

"NIST 800-53 is comprehensive but might be overwhelming for our immediate needs," Gary added, referring to one of many National Institute of Standards and Technology cybersecurity frameworks. "With over 1000 controls, it would be a significant undertaking to implement all at once."

Fran, who had been reviewing the cost data, looked up. "What about the financial aspects? Security is obviously critical, but our cost optimization work is also addressing significant risks. How does that fit into this framework?"

"That's an excellent point," Gary said. "This needs to be broader than just security."

"The FinOps Foundation has a framework that could help structure our financial controls," Fran suggested. "It's not as prescriptive as CIS, but it provides a maturity model for cloud financial management that aligns with our cost optimization initiatives."

"So we'd be looking at multiple frameworks for different aspects of cloud management?" Nancy asked.

"Initially, yes," Gary replied. "CIS for security controls because it's concrete and immediately applicable. The FinOps Foundation framework for financial management best practices. And we'd need to develop operational standards that align with our specific requirements."

"I think CIS makes sense as our security foundation," Adam suggested. "It's widely adopted, prescriptive, and gives us something concrete to implement right away."

"The CIS security controls would be a solid starting point," Gary agreed. "We could map the latest CIS Benchmarks for AWS, Azure, and so on to our IAPPs for audit purposes, which would satisfy Isabel's immediate concerns about security controls."

"And we can organize our existing cost optimization work aligned to the principles and KPIs from the FinOps Foundation framework," Fran added. "That gives us a structured approach to financial management that complements the security controls."

"So we use CIS as our security baseline, FinOps principles for cost management, and our safe landing zones as part of our operational standards that address our specific requirements," Charles summarized.

The committee spent the next hour developing their approach and focusing on reviewing the CIS Benchmarks and how they could map them to the IAPPs.

"We should present this first to the Cloud Champions network," Gary suggested. "Get their feedback before a broader rollout. And emphasize how this simplifies rather than adds to their existing responsibilities."

"And show how it directly addresses audit findings," Sarah added. "Many teams are struggling with those same gaps Isabel identified. This internal cloud framework we are developing gives them a more straightforward path to closing them."

"We also need to be explicit about implementation timelines," Charles noted. "Isabel specifically mentioned the need for reasonable expectations given the complexity involved."

Gary spent the rest of the day preparing for the cloud framework presentation. This would be the first real test of their approach, revealing whether teams would see it as enabling or constraining, as simplifying or complicating.

Day 278: Framework Revealed

Gary arrived early to set up for the cloud framework presentation. Just three days after the CEC had decided on their approach, this would be their first opportunity to share it with the broader organization.

The meeting included the Cloud Excellence Committee members along with the Cloud Champions, those key individuals from each division who had emerged as natural advocates for cloud best practices. If they could gain support from this group, broader adoption would be much more likely.

"Over the past several months, our team has been reacting to incidents, trying different recommendations, and learning through trial and error," Gary began once everyone had settled. "We've formalized some of this guidance into our landing zones as part of the Account Gateway Project, but until now, we haven't had a structured approach to controls."

He transitioned to the core of the presentation. "Following our recent discussions with Audit, we need to start formalizing how we're controlling cloud workloads, specifically how we're adhering to our IAPPs. The challenge is that IAPPs provide general control guidance not specific to the cloud, and there are too many controls to take on all at once."

Gary displayed the framework structure on the screen. "Rather than using IAPPs as our starting point, we're first adopting industry standards with prescriptive guidance for each cloud environment. CIS provides an overall control framework with specific benchmarks like the AWS Foundations Benchmark covering IAM, EC2, S3, and other core services."

"It's not all-encompassing across the hundreds of AWS services," he acknowledged, "but it's a practical starting point for security posture. We can implement these recommendations and map them to our IAPPs so we have a common language internally, especially with our Audit team."

He showed a slide with the specific CIS recommendation, the corresponding IAPP requirement, and the implementation approach for different cloud providers.

"For instance, the AWS CIS Benchmark in section 1 recommends ensuring IAM password policies require at least one uppercase letter. This directly satisfies IAPP requirement AC-12, which mandates complex password policies."

"What you're seeing is a translation layer that connects what we're already doing in the cloud to what Audit expects to see in their assessments," Gary explained. "Instead of creating new requirements, we're showing how industry best practices satisfy our existing policies."

The hands shot up immediately, with questions that revealed both interest and concern.

"How do you expect teams at different stages of cloud adoption to implement all these controls?" asked a Cloud Champion from Wealth Management. "Some of us are just starting our cloud journey, while others have sophisticated operations."

This was the opening Gary had been waiting for to introduce the maturity model approach.

"That's exactly why we've defined different implementation levels," he explained. "We're not expecting everyone to be at the advanced level immediately. This is iterative. We cannot assume full coverage overnight. Instead, we're providing incremental targets that reflect your team's specific cloud journey."

"For security, we've identified 'foundational' CIS controls that all production environments should implement immediately," Sarah added. "These focus on critical areas like identity management and public access prevention. The rest can be implemented on a longer timeline based on risk."

"So how will this be scored for adherence?" Jennifer asked. "Will all teams be expected to meet the same level across all controls?"

"We're still developing the scoring approach," Gary replied, "but our initial thinking is that production environments would be expected to meet at least the basic CIS level for all controls, while development environments would have different expectations."

Mike raised his hand. "This looks comprehensive, but how do we actually implement all these controls, especially for existing environments?"

"Great question," Gary responded. "This is where our implementation approach becomes important. We're leveraging our monitoring tools that already support CIS Benchmarks to identify compliance gaps."

Sarah picked up the thread. "For example, our security platform already has built-in checks for CIS controls. For AWS accounts, it can automatically verify if CloudTrail is enabled across all regions, which satisfies both an AWS CIS Benchmark recommendation and the IAPP requirement AU-03 for audit logging."

"And for new accounts, we're expanding our safe landing zones to automatically implement these controls from the start," Adam added.

As the discussion continued, Gary noticed that teams weren't debating whether the controls were necessary; they were focusing on how to implement them effectively.

"What I see as most valuable here isn't just the mapping to IAPPs," Peter noted, "but the implementation guidance. Telling teams 'enable multi-factor authentication' is easy, showing them exactly how to configure it across different services with specific CLI commands or console steps is what actually drives adoption."

"That's precisely our focus," Gary agreed. "Moving from defining what teams should do to enabling how they can do it efficiently."

After the formal presentation, teams broke into domain-specific discussions to explore sections of the CIS framework in more detail. Gary circulated through the groups, observing the technical and implementation-focused conversations. Even teams that had previously been resistant were engaging constructively with the specifics.

As the session wrapped up, Charles took a moment to observe the engagement level.

"This is very different from your town hall six months ago," he noted with a smile. "The focus on practical implementation seems to be resonating."

"We've learned a lot about how to introduce change in this organization," Gary replied. "But I think the key difference is that we're now bringing solutions, not just requirements. The framework isn't just telling teams what they need to do; it's showing them how to do it and why it matters."

Walking out of the meeting room, Gary felt a sense of cautious optimism. The framework presentation had gone better than he'd dared hope. Instead of resistance, they'd received engagement. Teams were treating this as a tool to help them succeed, not a burden to be endured.

Day 285: Cost Control

"Cost optimization is just as critical as security," Fran emphasized, presenting to the smaller group that had assembled for the financial management session.

One week after the presentation on CIS security controls, Fran had organized a focused workshop on cloud financial management. While security controls had understandably taken priority following Isabel's audit findings, the financial aspects of cloud usage couldn't be neglected.

"Our cloud costs have increased 42% year-over-year, while our actual workload volume has grown only 18%," Fran continued, displaying a chart showing Goliath's cloud spending trends.

The room included financial analysts from several divisions, IT budget managers, and a few Cloud Champions with budget responsibilities. Gary sat near the back, there to support Fran's initiative but letting her lead the financial discussion.

"Until now, our cost management has been reactive and fragmented," Fran explained. "Each division tracks spending differently, employs different optimization tactics, and follows different approval processes."

She introduced the FinOps Foundation framework, an industry standard for cloud financial operations.

"The FinOps Foundation provides a structured approach across three domains: Inform, Optimize, and Operate," Fran explained. "We start by providing visibility into costs and establishing ownership. Then we optimize through rightsizing and reserved capacity. Finally, we implement policies and continuous monitoring."

Dennis, who had been reviewing his division's latest cloud bill, spoke up. "We've been burned by unexpected costs several times. Developers spin up resources for testing, forget about them, and suddenly we're over budget."

"That's exactly the kind of challenge FinOps addresses," Fran replied. "For example, one core capability is establishing related cost allocation through consistent tagging."

She pointed to a slide showing resource tagging standards. "Building on the tagging work we started months ago, we're now extending those standards to enable financial reporting. The 'environment,' 'owner,' and 'application' tags we previously established provide the foundation we need for proper cost allocation."

She showed a screenshot of their dashboard highlighting cost visibility. "With proper tagging, we can track expenses down to the team level, enabling accountability."

"How does this connect to the security controls?" asked Jennifer.

"They're complementary," Fran explained. "Just as we're using CIS Benchmarks for security, the FinOps framework gives us structure for financial management."

She displayed examples of FinOps practices and their impact.

"One team reduced their EC2 costs by 72% by implementing automated instance scheduling for development environments," she explained. "Their non-production resources now automatically shut down outside of business hours."

Another slide showed the impact of reserved capacity purchases. "By analyzing usage patterns and committing to reserved instances for stable workloads, we've reduced the baseline infrastructure costs by 38% in pilot accounts."

"For implementation, we'll focus first on a few core practices," Fran continued. "Consistent tagging, budget alerts, and rightsizing recommendations. These will integrate with our Account Gateway process."

After the attendees had departed, Gary shared his observations with Fran.

"I noticed something interesting in there," he said. "When you showed the cost savings examples, people were immediately engaged. Teams seem to connect with financial improvements on a practical level rather than just seeing them as compliance requirements."

"That's been my experience too," Fran replied. "The FinOps framework gives us structure on the financial side, just as CIS does for security. Both are essential for effective cloud management."

Gary nodded. "You've made a strong case for the financial aspects. That's exactly what we need, a holistic approach that considers all dimensions of cloud usage. While Audit is immediately concerned about security controls, financial risk is also critical to demonstrate control. Unmanaged cloud spending can be just as significant a risk to the organization as security vulnerabilities."

Day 290: Barry's Shift

Barry studied the CIS controls mapping document on his screen, brow furrowed in concentration. Five days after Fran's financial management presentation, Dennis had requested this meeting to discuss Commercial Banking's approach to the security controls.

"So what's your assessment?" Barry asked, gesturing to the document.

"It's more focused than I expected," Dennis replied. "Using CIS Benchmarks and mapping them to IAPPs addresses many security challenges we've been facing, particularly around providing actionable guidance for different cloud services."

"And the phased approach makes sense," added Mike, Dennis's technical lead. "We can prioritize our efforts rather than implementing everything at once."

Barry nodded thoughtfully. This wasn't the rigid central mandate he had feared. The framework acknowledged different starting points, provided implementation flexibility, and focused on practical guidance rather than theoretical requirements.

"What about our existing security practices?" Barry asked. "We've invested significantly in our own controls since the incident."

"That's where it gets interesting," Dennis said. "The approach isn't dismissing what we've built. It's providing a structure to evaluate our controls against a common baseline. Where we exceed requirements, we continue our implementation. Where we have gaps, we get actionable guidance."

"But how will central teams verify our implementation if we're doing things differently?" Barry asked. "Won't this create more manual review work?"

"Sarah's security platform already has built-in checks for CIS Benchmarks," Dennis replied. "It examines our runtime configurations and can automatically verify if controls are implemented, regardless of our specific approach. It verifies the outcome, not the path we took."

Mike nodded. "Though I am concerned about some of our custom security implementations. The automated checks might not recognize them correctly, which could lead to false negatives in the reports."

"That's exactly why I'm asking about this," Barry said. "I don't want us flagged for non-compliance when we've actually addressed the risks in our own way."

As they discussed the implications further, Barry began to see something he hadn't initially considered. By aligning with the CIS Benchmarks as defined by the CEC, the responsibility for determining adequate security controls would shift from Commercial Banking to the central team.

"So if we adopt these benchmarks as our baseline," Barry mused, "and Isabel from Audit comes asking questions, she'd be primarily discussing control adequacy with the CEC rather than with us directly?"

"That's right," Dennis confirmed. "They own the control definitions and the mapping to IAPPs. We'd just need to demonstrate compliance with the benchmarks, not justify why our controls are sufficient."

Barry nodded slowly, the implications becoming more apparent to him. "And when exceptions are needed, the CEC would be responsible for documenting and defending them to Audit?"

"As long as we have legitimate business justifications, yes," Dennis nodded. "They're positioning themselves as the interface between divisional implementations and audit requirements."

A slight smile crossed Barry's face. Commercial Banking's strength had always been its focus on business innovation. The time they spent defending their approach and explaining why they were different was time not spent on delivering features. If the CEC wanted to take ownership of the security baseline and its justification to Audit, that actually worked in his favor.

"I'll discuss this with Charles," Barry decided. "But I'm inclined to have us commit to the CIS Benchmarks as our security baseline. We'll provide feedback on the implementation guidance, especially where our custom approaches need to be recognized, but we'll let the CEC own the framework itself."

The following day, Barry met with Charles to discuss his decision.

"Dennis shared the latest progress from the CEC with their new security baseline," Barry stated. "Commercial Banking would like to be one of the organizations compliant with these controls. Dennis and Mike will work with Gary's team to address any gaps and document necessary exceptions."

Charles couldn't hide his surprise. "That's a change from your usual approach to central initiatives."

"The approach addresses real challenges we've been facing," Barry acknowledged. "And as the CEC has continued to build their maturity around security and financial controls without forcing specific implementations on us, we've maintained our autonomy to get things done."

What Barry didn't mention was his realization that by aligning with the CEC-defined baseline, he was effectively transferring ownership of audit compliance to the central team while maintaining control over implementation details. It was a strategic repositioning that served Commercial Banking's interests while appearing collaborative.

Commercial Banking had been the most consistent source of resistance throughout Gary's time at Goliath. What looked like a significant shift in organizational dynamics was actually a savvy move by Barry to find a way to be cooperative while protecting his team's autonomy.

Day 302: Building Momentum

"We've got a problem, but it's a good problem," Gary said, looking up from his laptop during the weekly CEC meeting. "The CIS implementation assessment requests are exceeding our capacity to support them."

"It's not just the divisional IT teams," Sarah noted. "We're getting requests from individual application teams for guidance on specific controls. They want personalized advice on implementing benchmarks in their particular environments."

"The approach we took with Risk Analytics and Enterprise Data Services is working well," Gary observed. "Being active in architecture reviews and providing implementation support has shown real value. Now other teams want that same level of hands-on guidance for applying the security controls to their environments."

"The personalized guidance is what they're really after," Isaac added. "Teams understand they need to implement controls like cross-account access patterns, but they want help applying those patterns to their specific architectures."

This engagement was exactly what they had hoped for, but it created practical challenges. The central team simply didn't have enough resources to conduct deep-dive assessment sessions with every team that requested one.

"We need to scale our approach," Gary said. "We can't personally support every team through this process."

"What if we created a more structured self-assessment approach?" Adam suggested. "Something that builds on the dashboard but provides more context and guidance for teams to evaluate themselves before requesting our direct involvement."

"How would that be different from the dashboard we already have?" Nancy asked. "Or the alerts from our security and FinOps tools?"

"The dashboard shows metrics and compliance status," Gary explained, "but it doesn't guide teams through evaluating how they meet the control objectives or provide recommendations for improvement. The security and FinOps tools alert on specific issues but don't connect them to the broader framework."

"We need something that connects all these dots," Sarah added. "The dashboard tells you where you stand. The security tools flag specific vulnerabilities. But teams need guidance that ties these together, explaining which controls satisfy which requirements, how they relate to each other, and recommendations for implementation."

"A self-assessment tool could include guidance questions, examples of what 'good' looks like, and recommendations tailored to common scenarios," Isaac suggested. "Teams could use it to identify where they need the most help before engaging with us."

"And we could prioritize our direct engagement based on risk and impact," Sarah proposed. "Focus our limited resources on production environments with customer data or critical business functions."

The discussion continued, identifying ways to scale their approach while maintaining quality and consistency. By the end of the meeting, they had outlined a strategy that combined enhanced self-assessment tools, prioritized direct engagement, and expanded implementation guidance.

"There's something else we should consider," Charles said, joining the conversation. "The security and financial frameworks have momentum now, but sustaining that will require organizational structure. We need to think about how these efforts fit into our broader operating model."

This was a crucial consideration. What had begun as a response to the Jason security incident had evolved through security controls, account consolidation, and financial management frameworks. But as these initiatives became central to Goliath's cloud operations, they needed more formal support structures.

"Are you thinking about establishing a dedicated team to manage these frameworks?" Gary asked.

"Something like that," Charles confirmed. "A formal team with dedicated resources rather than just a committee that meets periodically. We'd need to define its structure, responsibilities, and how it works with existing teams."

The implications were significant. The cloud initiatives they had developed were transitioning from projects to ongoing capabilities.

Later that day, Gary met with Peter to discuss how the cloud framework was being received among the Cloud Champions network.

"Teams are engaging with it differently than past initiatives," Peter observed. "They're treating it as a tool to help them succeed, not an obligation to be managed."

"What's driving that difference?" Gary asked.

"The direct support," Peter replied without hesitation. "Previous standards told teams what to do but left them to figure out how. Now you're sitting with them, helping them understand how to apply the framework to their specific environments."

"We're also seeing teams help each other more," Peter continued. "When someone asks a question in the Cloud Champions channel, they get responses from peers before CEC members even have a chance to weigh in. What started as an informal network of advocates has evolved into a genuine community of practice. Division teams that never spoke to each other before are now sharing solutions and learning from each other's mistakes."

"We need to formalize this approach," Gary told Charles during their weekly update. "The collaboration between the CEC and Cloud Champions has been effective, but we need more sustainable structures as these initiatives become central to our operations."

"I agree," Charles replied. "Let's develop a proposal for how we might structure this work more formally within central IT. We can present it to the Executive Technology Committee next month."

This was the next logical step in their evolution, moving from project-based initiatives to a more permanent organizational capability.

Day 309: Cloud Team Evolution

Charles had called a meeting with the CEC members to discuss the team's growing responsibilities and organizational structure. It was time to elevate their approach and position.

"Thank you all for coming," Charles began once everyone had settled. "I've been reflecting on how far we've come since we first brought Gary on board to help mature our cloud operations."

He looked around the room at the assembled team. "When we hired Gary, we needed someone to make cloud a first-class citizen within central IT, to find patterns across our scattered workloads, and to bring some cohesion to our approach. A lot has progressed since then."

Charles continued, "We've had our challenges. From those initial attempts to rally business units around cost controls and tagging when they weren't ready for central IT's involvement, to security incidents that required a coordinated response. But look at what we've accomplished."

He began listing their achievements: "We've gained unprecedented visibility with our security and FinOps tools. We've built Account Gateway patterns and repeatable network and identity models in our safe landing zones. We've responded effectively to audit concerns. We've brought together central IT cloud efforts through this committee. We've helped business units like Risk

Analytics and Enterprise Data Services scale their workloads in the cloud. It's been an impressive journey."

Charles acknowledged each person in the room: "Each one of you: Sarah from Security, Fran from Finance, Nancy from Networking, Isaac from Identity, Adam from Asset Management, Oscar from Operations, has played a crucial role in maturing cloud at Goliath alongside Gary."

His tone became more serious. "We have momentum now. We have more business unit buy-in than ever before. We have the scrutiny of audit, which is actually a good thing, it means what we're doing is important enough to warrant attention. We've reached a point where we should elevate our work further, expand our scope, and fundamentally shift Goliath's cloud strategy to be governance forward."

Gary took over, outlining their current challenge. "We've grown beyond being an ad-hoc committee. Security controls, financial management, operational standards, and compliance requirements are all interconnected aspects of effective cloud governance."

"In practical terms, what changes?" Oscar asked.

"Three main things," Charles replied. "First, Gary would formally lead this function as Cloud Governance Director, reporting directly to me as part of my leadership team, rather than managing it from within Enterprise Architecture. This would essentially formalize the Cloud Team we've been building around Gary's work. Second, we'd establish dedicated resources focused on this work full-time, starting with our new cloud architect and potentially shifting more roles from on-premises teams as our cloud footprint grows. And third, the team would have defined organizational authority to drive standards adoption."

Adam looked thoughtful. "So when my team is developing asset management practices for cloud, we'd still own those practices, but we'd work through the Cloud Governance Team to implement them consistently?"

"Exactly," Gary confirmed. "We're not replacing your domain expertise or ownership. We're providing the coordination layer that makes implementation more effective."

Nancy seemed relieved. "That makes more sense. I was concerned we were creating a new control layer, but this is really about better supporting the work we're already doing together."

"Precisely," Charles affirmed. "Instead of this work being everyone's side job, it becomes a well-defined function with appropriate resources. This would be our new way of operating to ensure consistency and alignment on cloud best practices. The team would continuously refine our standards and patterns, but most importantly, help execute your processes across the organization with a focus on making the experience for application teams as agile and self-service as possible."

Fran nodded in understanding. "So the Cloud Champions network continues as our connection to the business units, the functional teams like mine continue to own our respective domains, and this new team formalizes the coordination role the CEC has been playing."

"That's it," Gary confirmed. "It's an evolution of what we have been doing, making it sustainable."

Gary paused, "And sustainable for the application teams. Our mantra should be enabling business agility, ensuring appropriate controls, and automating best practices. As a central function, we should be hellbent on helping teams move faster and safer in the cloud, or else we shouldn't do it."

"Exactly," Charles affirmed. "And with that approach guiding us, I believe we can transform how Goliath operates in the cloud."

Day 315: Conference Invitation

"The CloudNexis conference," Gary said, studying the invitation in Charles's office. "They want you to present on Goliath's cloud journey?"

Charles nodded. "They've been reaching out for years. I've always declined as we weren't ready to share our story. But after everything we've accomplished in the last few months..."

"A lot has changed," Gary agreed. "From struggling to identify shadow accounts to building a real cloud management approach."

"Which is why the timing of this works well," Charles said. "Our management formally approved your newly defined role as Cloud Governance Director, reporting directly to me. Real expectations now to build a formal shared service around cloud governance."

"And now a real deadline," Gary added, tapping the invitation. "Six months to publicly show what we can do."

Charles leaned forward in his chair, in a decisive tone said, “I want you to present instead of me.”

"Me?" Gary looked surprised.

"You've been at the center of this transformation. From those early cost optimization scripts to the Account Gateway to the security framework. You should be the one to tell our story."

Gary considered the opportunity. The past few months had been a journey from crisis management to a strategic approach. They'd built foundations, but still had much to do.

"This gives us a goal to work toward," Gary said. "We've made progress on visibility, security controls, and financial management. Now we need to bring it all together into something cohesive."

"Exactly," Charles replied. "It's not just about what we've done, it's about where we're going. CloudNexis is our chance to showcase how a traditional bank is handling cloud transformation."

Gary felt a sense of positive pressure forming. Six months to build on their foundations and create something worth sharing with the industry. Six months to define what cloud governance meant for Goliath in a practical, implementable way.

"I'll need to start getting feedback from others in the org on where they would like us to go," Gary said. "We've been working on separate tracks: security, finance, operations. We need to start thinking about how it all fits together."

Charles smiled. "That's exactly why I pushed for your role. Someone needs to connect those dots. The conference is just extra motivation."

As the meeting concluded, Gary's mind was already racing with next steps. Building on their successes while addressing the gaps. Creating a framework that brought together their various initiatives. Developing processes that worked in the real world, not just on paper.

What did "governance" even mean in the world of cloud?

Traditional governance conjured images of approval boards, slow processes, and restrictions. That approach wouldn't work in the cloud, where speed and flexibility were essential. They needed something different, a new definition that balanced control with enablement, security with innovation.

Before he could present their journey at CloudNexis, he needed to define what cloud governance meant for Goliath. Not as an abstract concept, but as a practical, value-driving capability that teams would actually benefit from.

Day 329: Alert Fatigue

"Teams are overwhelmed with alerts," Gary explained, spreading printouts across the conference table. "After talking with several application teams, I'm hearing the same feedback repeatedly."

Two weeks had passed since the CloudNexis invitation, during which Gary had been deeply engaged in defining his new role. He'd spent time with key business units and application teams, getting a more complete picture of their challenges with cloud adoption.

Today's meeting with Sarah, Adam, Fran, and Oscar was focused on a persistent problem: teams struggling to respond effectively to cloud issues despite having monitoring tools in place.

"Here's what I'm hearing from almost everyone," Gary continued, pointing to his notes. "Dennis's team receives security alerts directly from our CNAPP tool, cost recommendations from our FinOps platform, and operational metrics from multiple monitoring solutions. Everything comes through different channels, with different urgency levels, and no common context."

Sarah nodded in recognition. "Classic alert fatigue. They can't distinguish what's truly important from routine notifications."

"And it's made worse because different tools make different, sometimes contradictory recommendations," Gary added. "Financial optimization might suggest reducing instance sizes while performance monitoring indicates they need more capacity."

Fran nodded in agreement. "That matches what I'm seeing with cost optimization alerts. Teams appreciate knowing where they could save money, but they don't know how to prioritize those against security and operational concerns."

"The dashboard we created after the audit was supposed to help with this, right?" Adam asked. "Bringing everything into one view?"

"It was a good start," Gary acknowledged. "But it's become just another informational resource rather than a daily operational tool. Teams check it occasionally but struggle to translate what they see into specific actions."

Oscar nodded vigorously. "I've noticed the same thing. Teams understand their operational metrics in isolation, but they're not connecting them to security or cost considerations. Everything's in silos."

"It's become a reporting tool, not an action platform," Sarah observed. "We compile it manually, it provides a point-in-time snapshot, but teams aren't using it to drive remediation."

"This matches what I've observed with the Account Gateway," Adam added. "The gateway has helped with consistency for new accounts, but existing environments are still all over the place. Teams know they have issues but aren't systematically addressing them."

"What I'm seeing is a gap between awareness and action," Gary said. "Teams can see they have issues, but don't know what to do about them, which ones to prioritize, or how different alerts relate to each other."

"It's also a communication problem," Sarah suggested. "Each central team speaks their own language. Security talks about vulnerabilities, Finance talks about optimization, Operations talks about reliability. But application teams just want to know what they need to fix, in what order, and why."

Fran sighed in recognition. "They just want to know: what should I do today to improve my environment?"

Gary began sketching on his tablet. "We need to normalize this communication. Create a common language that connects all these different perspectives."

"And we need to help teams understand the context," Adam agreed. "An IAM finding might be critical in a production account, but lower priority in development. A cost optimization might be worth implementing in a long-running service but not in a temporary environment."

As the discussion progressed, Gary paused and looked around the table. "I think what we're talking about here is our end-to-end visibility of 'knowing our cloud'."

The team looked at him with interest.

"Think about it," Gary continued. "Before teams can improve their environments, they need to truly understand them. Not just isolated metrics, but a comprehensive view with proper context."

Oscar nodded slowly. "Resource inventory, operational health..."

"Security posture, cost profile..." Sarah added.

"Compliance status, ownership information..." Fran continued.

"Exactly," Gary said with growing excitement. "These are the fundamental building blocks of knowing your cloud. Our existing tools provide all this data, but in fragmented ways that don't connect to action."

"These tools can tell us what's happening," Adam noted, "but they can't manage the process of implementing changes across the organization. That's the gap."

"And they're not integrated with our governance framework," Sarah added. "We mapped CIS controls to our IAPPs, but there's no connection between those mappings and the alerts coming from our security tools."

"This feels like we've identified a major part of cloud governance," Gary observed. "'Know Your Cloud.' Having visibility is critical, but there's a next iteration of transforming that into intelligence. Our next challenge is taking what we know about cloud environments and communicating it in a way that enables teams to act effectively."

Day 336: Lost in Translation

"It looks like a technical document written by security engineers for other security engineers," Dennis said bluntly, sliding Gary's cloud framework document across the table. "My developers won't read past the first paragraph."

A week after identifying the visibility gap, Gary had attempted to translate their cloud control framework into implementation guidance for application teams. He had approached it like a standard operating procedure that was technical, comprehensive, and meticulously documented. The feedback was direct. The guidance wasn't connecting with the people who needed to use it.

"I appreciate the honesty," Gary replied. "What specifically isn't landing?"

Dennis flipped to a highlighted section. "Take this part on resource policies best practices. It's full of technical jargon and references to AWS documentation, but lacks concrete examples of what we should actually implement. My team doesn't need a lecture on the principle of least privilege. They need templates for common service patterns they can adapt to their use cases."

"We're talking at them, not with them," Gary observed.

"Exactly," Dennis agreed. "My team already gets alerts from your security tools, recommendations from the FinOps platform, and metrics from monitoring dashboards. But none of these connect back to your control framework in a way that helps them understand what to actually do."

The problem was emerging. The framework itself was sound, but the connection between framework controls and daily implementation was missing.

"The other issue is coordination," Dennis continued. "Security sends requirements, Finance sends cost recommendations, Operations sends reliability standards. Sometimes they contradict each other, and there's no single source of truth for what we should implement."

After Dennis left, Sarah joined Gary to review the feedback.

"This matches what we've been hearing from other teams," she said. "Our control framework addresses the right concerns, but the communication is fragmented. Different tools are sending different messages without a unified context."

"We built a framework of controls," Gary said, "but we haven't normalized how those controls are communicated to application teams."

Over the next few days, Gary worked with Sarah, Adam, and Fran to develop a new approach to communicating cloud controls. During a whiteboard session, they discussed what elements every control communication needed.

"First, we need a concise summary," Sarah suggested. "A name, identifier, and severity rating that immediately communicates what the control is and how important it is."

"And business context is crucial," Fran added. "Teams need to understand why the control matters in terms of risk, compliance, and application impact."

"Implementation guidance should be practical," Adam noted. "Step-by-step instructions with working examples that teams can actually follow."

"We should also include an impact preview," Gary contributed. "Teams want to know what would happen if they applied the control to their environment."

"And don't forget a straightforward exception process," Sarah concluded. "Teams need to know how to request and document legitimate exceptions."

"The tone is as important as the content," Fran observed. "We need clarity over completeness. Teams don't need to be impressed by our technical knowledge; they need to understand what to do."

"And we need to be human-centric," Adam added. "Instead of saying 'This resource policy violates best practices,' we could say 'This permission creates a risk that could potentially expose customer data.'"

By Thursday, they had developed a prototype communication for one of their critical controls: “No Public AWS EBS Snapshots.” Instead of a technical document, they created a concise overview with actionable guidance.

The new format started with a simple identifier, which IAPP control it was associated with, and severity rating, followed by a concise summary explaining that any EBS snapshot must be set to private. It explained why the control mattered; as EBS snapshots could contain confidential information, they are intended to be set to private by default, accessible only in the owner’s cloud account. It provided specific implementation steps, with supporting details for how to implement via the console, CLI, and Terraform. The communication ended with guidance on getting help or requesting exceptions.

"This is much more digestible," Sarah noted as they reviewed the prototype. "Concise, actionable, and gives context without overwhelming."

When Dennis reviewed the new approach with Mike from his team, their response confirmed they were on the right track. The format bridged the gap between the framework's technical requirements and the practical implementation details teams needed.

With the communication approach established, the team began converting their existing controls into the new format. The work was painstaking but energizing. They were finally bridging the gap between framework and implementation, between knowledge and action.

Day 343: Control Rollout

"Sending better emails isn't enough," Gary said, opening their follow-up meeting on control implementation. "We need a more structured approach to rolling out controls across the organization."

The team had gathered to build on their improved communication template. Dennis's feedback on their prototype had been positive, but highlighted a deeper challenge: even well-crafted guidance would fail without a thoughtful implementation strategy.

"I've been thinking about this in terms of organizational change management," Gary continued. "Each control we implement represents a change that teams need to absorb and adopt. We need to treat these rollouts with the same care we'd give any significant organizational change."

"That makes sense," Sarah replied. "We can't just flip a switch and expect teams to immediately comply with new controls. They need time to understand, prepare, and implement."

Adam, who had been sketching on his tablet, shared his screen. "What if we thought of each control as a deployment? A control rollout? Not just a one-time notification, but a structured process with distinct phases?"

"Like a kanban workflow approach," Fran suggested. "Moving controls through different stages of implementation across the organization."

Oscar, who had been listening quietly, spoke up. "From an operations perspective, this phased approach makes a lot of sense. It gives teams time to incorporate these changes into their planning rather than scrambling to react to sudden requirements."

This concept resonated with the team. Over the next hour, they outlined a phased rollout approach for cloud controls.

Gary sketched on the whiteboard as they developed the process.

They would start with a "Draft Phase" where the Cloud Team would internally test and validate each control.

Once refined, they would move to a "Preview Phase," sharing details with application teams to gather feedback before actual implementation.

Next would come the "Check Phase," where they would enable monitoring and alerting for the control but without enforcement. This would give teams visibility into their compliance status and self-service tools to remediate issues.

When applicable, they would eventually progress towards an "Enforce Phase" with automated remediation that would continuously maintain compliance and automatically correct any drift. But only after adequate preparation and testing.

This graduated approach would build trust while ensuring controls were eventually implemented.

"This gives teams time to adapt," Gary explained. "Instead of just turning on a control and watching alerts flood in, we're creating a collaborative process that progresses from team self-service to automated protection."

"And it gives us time to refine our approach based on feedback," Sarah added. "We might discover implementation challenges we hadn't anticipated."

Dennis joined them later that morning to review the proposed rollout approach.

"This addresses my biggest concern," he said after Gary walked through the phases. "In the past, we'd suddenly start getting security alerts about a new requirement with no warning. By the time we found out about it, we were already non-compliant."

"The preview phase is particularly valuable," Dennis continued. "Seeing the potential impact before the control goes live gives us time to plan our response, request exceptions if needed, or adjust our configurations proactively."

"And how would teams track where controls are in this process?" he asked.

Adam pulled up a mockup of a dashboard showing the status of various controls across the organization. "We're envisioning a central view where teams can see which controls are in which phase for their environments. They'd get notifications as controls move through the phases, with established timelines for each transition."

"We're also planning to add a feedback loop," Gary explained. "At each phase, teams can provide input that helps us refine both the control itself and our communication about it."

By midday, they had a solid framework for rolling out controls across the organization. But implementation questions remained.

"What about exceptions?" Sarah asked. "Not every control will apply equally to all teams. We need a way to handle legitimate business or technical reasons why a control can't be implemented."

"That's an important point," Gary agreed. "We'll need a structured exception process as part of this approach. Something that balances flexibility with accountability."

"We can outline the basics now," Adam suggested, "but we'll need to work through the details when we encounter our first real exception case."

As they wrapped up, Gary realized they were defining another part of cloud governance. "Last week, we focused on helping teams know their cloud through better visibility to 'Know Your Cloud.' Now we're establishing how to raise the bar through these control rollouts."

"Raise the Bar," Sarah repeated thoughtfully. "That's exactly what this is. Its defining processes for how we can take systematic action to improve our security posture, cost efficiency, and operational excellence."

"And doing it in a way teams can actually adopt," Adam added. Processes for how to "'Raise the Bar' have to be achievable, or teams will just work around it."

Day 350: Exception Dilemma

"Our team needs an exception for the 'No Public AWS EBS Snapshots' control," Dennis explained, pointing to the architecture diagram on his screen. "We have these snapshots shared with our disaster recovery provider, who operates in a separate AWS account."

Gary had convened a meeting with Dennis and his team to discuss their first formal exception request since implementing the snapshot restriction control rollout. While most applications had been able to comply without issue, Dennis's disaster recovery integration posed an edge case to consider.

"The security scan flags them as non-compliant because they're not private to our account," Dennis continued. "But this is intentional sharing with a trusted partner. It's how our disaster recovery system works, and it's not going to change."

"They're not public to everyone, just shared with a specific account?" Sarah asked from across the table.

"Exactly," Dennis confirmed. "They're shared with a specific account ID belonging to our DR provider for disaster recovery. The scan is technically correct that they're not private, but the sharing is intentional and secure."

Sarah nodded in understanding. "This highlights one of the challenges with security scanning tools. They see a binary state, either private or public, without the business context of why sharing might be appropriate in specific cases."

This was the governance dilemma in practice: balance security requirements against operational reality and risk level. The team was testing their rollout approach with a real-world scenario, and the manual process was proving cumbersome.

"Let me check our tracking sheet," Gary said, opening the spreadsheet where they were documenting the snapshot rollout. He filtered for FinTech accounts and reviewed the status. "Are all your other snapshots properly restricted except these specific DR ones?"

"Correct," Dennis replied. "Everything else is compliant. How do we document and track this exception to avoid talking about this again? Email approval? Update the spreadsheet? Does the request through the Cloud Portal need to be updated?"

"All of the above, apparently," Gary replied with a hint of frustration. "We're tracking this rollout across systems from the Cloud Portal requests, spreadsheets, and emails."

They spent the next thirty minutes discussing how to document the exception. Who needed to approve it? Where would it be recorded? How would they track the expiration date for revisiting the exception? Each question revealed another gap in their manual tracking process.

"At minimum, we need to document the specific requirement being excepted, the business justification, any compensating controls, and a review timeline," Gary suggested, also noting, "and a way for us to be aware not to send you the same communication again while you are in an exception situation."

"And who makes the final decision?" Dennis asked. "My team owns the application, Barry would be the ultimate functional approver, but Sarah's team owns the security controls, and ultimately it's about risk acceptance."

This opened a broader question about ownership and accountability. The account belonged to the application team, but security controls crossed organizational boundaries. Their manual tracking didn't have a systematic workflow for routing approvals or documenting decisions.

"For now, let's say the application owner documents the justification, Security reviews it, and both sign off," Gary proposed. "But we'll need a more formal process going forward."

As Gary and Sarah were walking out of the meeting, Gary, a bit worried, said, "This is just the beginning. We have dozens of controls to implement across hundreds of accounts. This manual approach won't scale."

Sarah said with some positivity, "The Preview Phase worked exactly as intended, engaging application teams early and identifying practical challenges before enforcement. Think about the organizational noise this would have caused with Barry and others if we didn't communicate, fired off the alerts, and just marked them red in the dashboards."

The snapshot security control rollout would continue, and they'd document the exception as best they could in their spreadsheets and emails. But it was becoming apparent that scaling their governance approach would require a more comprehensive governance platform to manage how they could "Raise the Bar" and actually make change happen across their organization.

Day 357: Remediation Roundabout

"I feel like I'm spending half my day just managing this spreadsheet," Gary muttered, scrolling through the massive tracking document that had become the center of his working life.

Seven days after their first major exception discussion, the “No Public AWS EBS Snapshots” rollout had moved to the "Check Phase" for several test accounts, with automated scanning and generating alerts for non-compliant buckets. What had seemed straightforward in theory was proving far more complex in practice and repeating where the team had been before the rollout process.

Gary toggled between the security dashboard and his tracking spreadsheet, manually reconciling the difference between what the tool reported and what they had documented. The security platform showed 47 non-compliant buckets, but 9 of those had approved exceptions that needed to be backed out of the metrics.

"The percentages look good," Gary acknowledged, addressing Sarah. "But I'm more concerned about the remediation patterns we're seeing."

On his screen, he pulled up a different tab of the tracking sheet, this one showing the status of alerts sent to different teams.

"Look at these response patterns," he said, highlighting several rows. "Peter's team remediates issues almost immediately. But most other teams are barely responding, even though our alerts now include better context and remediation guidance than we've ever provided before."

"I thought the improved communication templates would help," Sarah said, frowning. "Teams now have detailed step-by-step instructions on how to fix each issue, but they're still ignoring the alerts."

"I believe we significantly changed everyone's awareness and engagement," Gary replied. "But we haven't addressed their ability to prioritize and manage the overhead of fixing the problem. We're competing with feature development and other operational work."

As they analyzed the response patterns across teams, several distinct behaviors emerged. Peter's team, with their mature DevOps practices, stood out for their efficiency. They quickly remediated issues and immediately updated their deployment pipelines to prevent recurrence.

In stark contrast, some teams were simply ignoring alerts entirely, treating security notifications as background noise amid their countless other operational messages. Others would fix issues when directly prompted, only to have identical violations reappear days later when they created new resources without applying the same fixes.

Perhaps most frustrating were the teams that had properly fixed issues in their deployment pipelines, but then undermined their own progress when someone made manual changes through the cloud provider console, bypassing all their carefully implemented guardrails.

"This last one is particularly frustrating," Sarah noted, pointing to an example. "The Risk Analytics team updated their Terraform templates your team helped them with to make sure no snapshots had public configurations, but then someone went in via the console and created a new snapshot set to public that bypassed those controls."

"The roundabout effect," Gary sighed. "We're going in circles with some of these teams."

Sarah nodded, pulling up another example. "And even after we follow up, it takes weeks to get simple fixes implemented. I've had three reminder emails with this Enterprise Data Services team, and they keep promising to fix it 'soon.'"

The remediation tracking had become increasingly complex. Gary's spreadsheet now included columns for initial notification date, acknowledgment date, promised remediation date, actual remediation date, verification date, and recurrence tracking. Each alert required multiple touch points and follow-ups.

Their discussion was interrupted by a message from Robert in Risk Analytics: "Getting alerts about three new non-compliant buckets. Can your team just fix them for us? We don't have the bandwidth to keep chasing these down."

Sarah peered at the message over Gary's shoulder. "That's the third request this week asking if we can just fix issues for them."

"Ironically, that's exactly what I was trying to do when I first joined Goliath," Gary reflected. "Remember that cost optimization script I wrote? I wanted central IT to help fix issues, but most teams resisted. Now they're asking for it."

"It's a natural evolution," Sarah suggested. "Once teams understand the requirements better, many realize they'd rather have central services handle routine governance tasks so they can focus on their core work."

The message from Risk Analytics reminded Gary of their earlier landing zone discussions where teams had been asking for more pre-configured environments and standardized patterns. This is the same concept, but for ongoing governance controls.

Day 364: Enforcement Experiment

"How exactly would automated enforcement work with our current setup?" Sarah asked, studying the whiteboard where Gary had sketched out some initial ideas.

A week after identifying the remediation challenges, Gary, Sarah, and Oscar were brainstorming solutions to the growing alert fatigue problem. The manual remediation approach clearly wasn't scaling, and teams were increasingly asking for help.

"We've already defined the Enforce Phase in our rollout model," Gary replied, "but we need to determine how it works in practice. I'm thinking we could start with a simple script that automatically sets snapshots to private."

Sarah nodded slowly. "Our security tool has some capabilities to run a script we could leverage, or we could build something cloud-native with AWS Lambda functions."

"But we'd need to be careful about scaling and exception handling," Gary cautioned. "Writing a script for one control is straightforward, but maintaining thousands of scripts across dozens of control types would quickly become unmanageable. Let alone handling conditional logic and application nuances. For example, what if our automated remediation tried to 'fix' Dennis's purposely shared snapshot with their vendor?"

Gary was acutely aware of their resource constraints. His lone cloud architect, Mark, hired a couple of months ago, was already stretched thin, primarily helping business units implement landing zones and supporting architectural patterns. They didn't have the engineering staff or programming expertise to build out a comprehensive custom automation framework.

"What about different models for enforcement?" Sarah asked, adding to Gary's whiteboard sketch.

They spent the next hour discussing different remediation approaches they could offer teams.

"The fundamental question is one of responsibility," Gary said. "Application teams own their environments and are ultimately responsible for compliance. Our role isn't to fix things for them but to enable them to fix issues more efficiently."

Sarah nodded in agreement. "We definitely don't want teams asking us to manually fix issues for them. That's not sustainable and undermines the ownership model."

"The challenge is that teams know they should fix these issues," Gary added. "With our improved communication templates, they understand what to do and why. But the reality is they get busy, other priorities emerge, and security fixes fall to the bottom of the backlog."

"Exactly," Oscar agreed. "We expect them to fix the issues, but we need to be more mindful in our approach. The endless back-and-forth of 'please close this alert' emails is consuming everyone's time."

"What about self-service remediation?" Adam suggested. "We could provide one-click tools that teams can trigger themselves when they're ready to fix an issue."

"That's a practical first step," Gary agreed. "It maintains their control while making compliance easier."

The conversation turned to more advanced automation options. Oscar proposed a delayed enforcement model where teams could opt into automated remediation that would trigger only after alerts had been ignored for a defined period.

"This gives teams a chance to fix issues themselves first," Oscar explained. "But ensures things don't remain broken indefinitely."

"And once a resource becomes compliant," Sarah added, "we could keep it that way through continuous enforcement that prevents future drift, regardless of whether the initial fix was manual or automated."

They considered more invasive approaches like pipeline scanning and preventative enforcement through IAM boundaries, but decided to defer those for future consideration.

"For now, we'll focus on self-service remediation and runtime enforcement options," Gary said. "Pipeline integration and preventative controls can come later when we see more demand from teams."

"For each model, we'll need to think about exception handling," Sarah added. "How do we ensure automated remediation doesn't affect resources with approved exceptions?"

"There's also the question of verification," Gary noted. "We need to confirm that automated remediations were successful and didn't cause unintended consequences."

"And what happens when a remediation fails?" Sarah added. "We need error handling, notifications, and retry mechanisms."

By the end of the day, they had a design for their enforcement experiment, but also a growing list of concerns about scaling the approach. Their plan was to start small, implementing automated enforcement for setting EBS snapshots to private on a few volunteer accounts from the Commercial Banking, Risk Analytics, and Enterprise Data Services teams.

"I keep thinking about how the teams' attitudes have changed," Gary reflected as they wrapped up. "Six months ago, they would have strongly resisted any form of automated enforcement. Now they're asking for it."

"It's about trust and value," Sarah suggested. "They've experienced the pain of manual governance and want something better. Making this change happen will require sustainable processes that work with Goliath's culture, not against it."

Day 371: Risk Analytics Opportunity

"We'd like to build a new fraud detection system using cloud-based AI and machine learning services," Robert explained, spreading architecture diagrams across the conference table. "But we need help making sure we do it right from the start."

"Tell me more about what you're trying to accomplish," Gary said, studying the diagrams.

Robert, the Risk Analytics director, explained their vision, "This would consist of a machine learning system that could analyze transaction patterns to identify potential fraud in real-time. We believe we need to leverage additional cloud native AI and ML services like AWS Bedrock, SageMaker and Comprehend, but we're concerned about security, compliance, and staying within budget."

"We've really enjoyed the partnership we have, with your team working collaboratively on our other cloud accounts." Robert continued. "We'd like to implement work together to architect something similar for this project, but with specific protections for AI workloads that are new to us."

Gary glanced at Mark, the cloud architect who had joined his team two months ago. This was exactly the kind of opportunity they'd been hoping for. A chance to implement governance from the beginning rather than retrofitting it later.

"AI is new for the organization, and specifically for us in the Cloud Team, we have not developed any architecture or governance patterns for these types of workloads. However, this is a great opportunity for us as an initial pilot that could be shared with others in the organization," Gary replied.

Over the next hour, Robert's team walked through their requirements. The application would process large volumes of transaction data, some containing sensitive customer information. They needed to ensure appropriate controls around data access, model training, and inference endpoints.

As the Risk Analytics team left, Gary and Mark remained to discuss the opportunity.

"This could be an important pilot," Gary said. "Not just for implementing controls from the start, but for developing AI-specific governance practices."

"I've been researching this area," Mark replied, pulling up some notes on his laptop. "There are several dimensions we need to consider for AI workloads that go beyond our standard controls."

As the discussion progressed, they identified several key areas for AI governance that needed to be addressed.

"We should start with model approval and inventory," Mark suggested. "Tracking which AI models are approved for use gives us immediate visibility and control over what's being deployed."

Gary nodded in agreement. "That's practical and impactful. We can implement a registry of approved models relatively quickly."

"And data protection has to be our other initial focus," Gary continued. "I'm sure Sarah would emphasize this point to ensure sensitive data used for training is properly secured. It's non-negotiable, especially in banking."

"Let's start with just these two controls rather than trying to tackle everything at once," Gary decided. "Model inventory gives us visibility, while data protection addresses our most critical risk. We can iterate and expand from there."

"We could develop these as an extension to our standard landing zone patterns," Mark suggested. "A specialized template for AI workloads."

"And we could implement continuous enforcement from the beginning," Gary added. "Instead of going through all the phases, we could establish the controls upfront and have them in an 'Enforce Phase' from the start."

"That would help reduce overhead from RA and our team," Mark noted. "The automation would handle remediation automatically."

The next day, they reconvened with the Risk Analytics team. When they presented their approach, the reception was enthusiastic.

"This is exactly what we were hoping for," Robert said. "We get security and compliance built in from the start, without slowing down our development."

As the Risk Analytics team left, Gary felt a sense of validation. This was governance as it should be, enabling rather than restricting, collaborating rather than dictating.

"I am feeling good about this engagement, but the challenge will be scaling those additional controls we plan on automating," Gary noted. "Our current spreadsheet tracking and manual implementation won't support enterprise-wide adoption."

Mark nodded in agreement. "The approach works. Now we need tools that match the sophistication of what we're trying to accomplish."

Day 378: Bridging Visibility to Action

"So that's our current process," Gary concluded, gesturing to the whiteboard filled with arrows connecting various tools, spreadsheets, and manual steps. "It's functional, but it isn't scalable."

Gary had called a working session with the core members from the Cloud Excellence Committee to evaluate their cloud governance approach. Sarah from Security, Fran from Finance, Adam from Asset Management, Oscar from Operations, Nancy from Networking, and Isaac from Identity Management gathered around the table, studying the increasingly complex diagram.

Gary had mapped out their governance ecosystem on the conference room whiteboard. The resulting diagram looked like a complex subway map, with lines connecting various components: Account Gateway landing zones, security controls mapped to CIS Benchmarks and their IAPPs, FinOps cost optimization recommendations, network architecture patterns, identity management guidelines, operational best practices, custom Excel-based dashboards, and the latest control rollouts they were piloting. Between these components were dozens of manual handoffs, spreadsheet tracking systems, scripts, and ad-hoc communications.

"We've built some impressive capabilities over the past year," Gary observed, "but they're all loosely connected through spreadsheets, basic dashboards, and manual processes."

Sarah nodded. "From the security side, our CNAPP tool gives us excellent visibility into vulnerabilities and misconfigurations. We can detect issues across hundreds of accounts efficiently." She pointed to another section of the diagram. "But translating those findings into actionable tasks for application teams, tracking remediation progress, and verifying fixes, that's all manual work spread across different systems and teams."

"Same challenge with cloud costs," Fran added, opening her laptop to show a recent report. "Our FinOps platform identifies optimization opportunities worth millions annually. But we have no systematic way to assign these to teams, track implementation, or measure outcomes. Each recommendation becomes its own mini-project."

Oscar gestured to the operations section. "And here's where we track reliability metrics, service level agreements or SLAs, and operational best practices. Again, great visibility into the issues, but no connection to our governance processes for driving improvements."

"Our Account Gateway has been a success for new cloud accounts," Adam noted. "Teams get standardized landing zones with built-in governance controls. But we have no way to detect when those environments drift from our baselines or to automatically apply new controls as they're developed."

Gary studied the diagram thoughtfully. "What I'm seeing is that we've gotten really good at understanding our cloud environments - identifying issues across security, cost, and operational dimensions."

"That's right," Nancy agreed. "Our visibility layer is solid. We know what's happening in our cloud environments from security, cost, and operational perspectives. But that visibility isn't translating efficiently into action."

"And we've made progress on defining systematic approaches," Gary continued, pointing to their control framework documentation. "We know how to structure controls, map them to standards, and create rollout phases. But there's still a gap between having good processes and actually executing them at scale."

He traced his finger along the arrows connecting different systems on the whiteboard. "We're struggling with the execution across all these teams and environments. How do we coordinate between security alerts, cost recommendations, and operational requirements?"

Isaac pointed to the identity section of the diagram. "Take IAM policies as an example. We have standard patterns defined in our landing zones. We can detect when teams deviate from those patterns, but there's no streamlined process to guide teams through remediation when misconfigurations occur."

"And the connection between these stages is entirely manual," Sarah added. "We'll identify a security vulnerability through our CNAPP tool, alert an application team, and the Cloud Team is chasing them down off to the side while reporting on how the controls are impacted. Repeated back-and-forth emails to manually remediate, review exceptions, and escalations to management take weeks, months, or never completed."

"We need a way to help us bring together this visibility to action," Gary summarized. "Let's spend the next part of the meeting outlining requirements for what a cloud governance platform would need to offer."

The team spent the next hour detailing what they needed in a platform.

"At its core," Gary began, sketching on the whiteboard, "we need to centralize security, cost, and operations recommendations from all our different tools. Right now, that information is scattered across systems."

"And we need to translate those recommendations into pre-built, extensible controls," Adam added. "Controls that can be mapped to our internal standards, our cloud framework."

Oscar nodded in agreement. "We also need workflow capabilities to move controls through implementation phases across all our cloud accounts. The rollout approach we've been developing manually needs to be built into the platform."

"Communication is key," Fran pointed out. "We need extensible templates and workflows for notifying teams about controls in a consistent way."

As the discussion continued, they identified additional essential capabilities: stakeholder and approver assignments for different controls, exception management workflows, landing zone deployment with built-in governance checks, and automated enforcement and remediation capabilities.

"The user experience is just as important as the technical features," Gary emphasized. "Application teams need self-service remediation options that make compliance simpler."

"And self-service visibility," Nancy added. "Teams should be able to see their control status and compliance posture without having to ask us."

They discussed other user experience requirements: simplified communication about new requirements and alerts, flexible communication channels between application teams and the Cloud Team for feedback and exceptions, and configurable rollouts and communications that could be tailored to different teams' needs.

"This platform would need to work for both sides," Gary emphasized. "Making governance easier for our Cloud Team while also making compliance easier for application teams."

Fran looked concerned about the budget implications. "Management will likely ask how this is different from the tools we already have. They'll see it as just another expense."

"What I'm starting to see," Gary replied, drawing on the whiteboard, "is that our governance journey has three distinct challenges."

He drew the first element. "First, you need to 'Know Your Cloud' - understanding what you have. Our current tools excel at this visibility layer."

He added the second element. "Then there's defining systematic approaches and building the organizational capability to support them, 'Raise the Bar.' We've made good progress here with defining our control frameworks, rollout phases, standards mapping, and getting the right people and processes in place through the Cloud Excellence Committee. We know what good looks like, how to structure improvements, and we have the team alignment to drive it forward."

Gary paused, studying what he'd drawn, then added the third element. "But the gap we're discussing today is actually executing those standards across hundreds of accounts and dozens of teams. Taking our visibility and our defined processes and making the organizational change happen at scale."

He stepped back from his diagram. "That's really what we're trying to solve, 'Make Change Happen.' We need something that bridges these gaps and helps us orchestrate execution across the organization. It's exactly why we formed the Cloud Governance Team in the first place."

Fran studied the requirements they'd outlined. "Do platforms like this even exist? This seems very specific to how we've evolved our approach."

"That's what we need to find out," Gary decided. "Let's evaluate what cloud governance platforms might meet our requirements and see if any align with our approach."

Day 385: Platform Exploration

"Let me summarize what we've found," Gary said, gesturing to the whiteboard where he'd organized their findings. He and Mark had spent the week researching cloud governance platform options, with input from Sarah and Fran on specific requirements.

"We evaluated five distinct categories of solutions," Gary continued, writing each on the whiteboard.

Developer-Centric & Open-Source Tools

"Several platforms, both commercial and open-source, promise complete flexibility through code-based policy engines," Mark explained. "They're powerful, offering customization through infrastructure-as-code approaches."

Gary nodded. "But they essentially require us to build and maintain a governance platform ourselves. They provide a foundation, but everything else, like policy definitions, remediation logic, exception workflows, reporting dashboards, becomes custom code we'd have to maintain."

"Given our engineering constraints, that's not sustainable," Mark added. "We'd be trading spreadsheet maintenance for code maintenance. And most lack critical features like exception management and communication workflows, we would still be managing manually."

Cloud-Native Services

"Each cloud provider offers native governance services," Gary noted. "They're well-integrated with their respective platforms and offer some automation capabilities."

"The obvious limitation is that they're cloud-specific," Mark replied. "AWS has their solution, Azure has theirs, GCP has something different. We'd need separate processes for each cloud."

"And they typically cover only a fraction of the available services," Gary added. "They focus on popular services but leave gaps in coverage for specialized services or new offerings. Additionally, there's no communication workflow, which we've identified as a critical requirement."

Traditional Service Management and Enterprise GRC Tools

"Several service management, or enterprise governance, risk, and compliance platforms claim to handle cloud," Gary explained. "They're designed for traditional IT governance with cloud modules added on."

"But they fundamentally misunderstand cloud-native operations, and are very much tailored to on-premises infrastructure," Mark countered. "Not able to support the real-time nature of cloud environments with the breadth and depth we are looking for."

CSPM, CNAPP, FinOps Tools

"This was interesting," Sarah noted, joining the discussion. "Several security and FinOps vendors are positioning their tools as 'automated remediation,' but they're primarily focused on finding issues, not managing remediation."

"Exactly," Gary agreed. "They aggregate alerts from Cloud Security Posture Management tools or features, and FinOps tools, providing nice dashboards showing findings across environments. They may have some action workflows you can enable. But they lack the workflow capabilities for systematically implementing controls, managing exceptions, and driving remediation."

"And their pricing models are concerning," Fran added. "Most require substantial upfront commitments based on cloud spend or resource counts, with unpredictable scaling costs as we expand our coverage."

"Plus, we already know from our existing tooling that we can't solve what we are trying to accomplish with them."

Cloud Governance Platforms

"Finally, we found a small category of purpose-built cloud governance platforms," Gary concluded. "These are designed specifically for the challenge we're facing, to bridge the gap between visibility and action across multi-cloud environments. Some look more developer-centric, others more around visibility and less around making change happen."

"This one stands out," Mark said, pointing to their evaluation matrix. "It offers the best balance of technical capabilities and governance workflows. Its control library already aligns well with the CIS Benchmarks we've mapped to our IAPPs, and it supports a control rollout deployment feature with multiple enforcement models: alerting, self-service remediation for app teams, and auto-remediation for continuous compliance."

"It doesn't have every feature we've identified," Gary acknowledged, "but looking at their release cadence, they're constantly iterating and have the broadest coverage we've seen."

"And importantly," Fran noted, "they offer consumption-based pricing. We can start with a free trial and transition to paying only for what we use. No large upfront commitment."

"The initial implementation would be manageable too," Mark added. "We could start with a specific domain like our next security control, or get started with a cost optimization we have been planning on. We can start small, prove the value, then expand."

Fran, who had been running numbers on her laptop, looked up. "The financial model looks promising, but we'll need a solid business case. How do we justify another platform when we already have investments in CNAPP, FinOps, and operational monitoring tools?"

"That's our next step," Gary agreed. "We need to quantify the benefits with time saved from manual processes, faster remediation, better resource utilization, and improved compliance posture. Let's build that business case and review it with Charles next week."

Day 392: Business Case

"Automating the cloud governance process sounds promising," Charles said, reviewing the presentation Gary had prepared. "But help me understand why we can't accomplish this with our existing tools and processes."

Gary, Mark, Fran, and Sarah had gathered in Charles's office to present their business case for a cloud governance platform.

"Our cloud adoption has accelerated faster than we anticipated," Gary began. "Mark is spending most of his time helping teams onboard to our landing zones, while I'm trying to manage the growing backlog of governance issues our visibility tools are identifying."

Gary displayed a dashboard showing alerts from their monitoring tools. "Our CNAPP tool, FinOps platform, and operations monitoring are excellent at finding issues. The problem is what happens next, or more accurately, what doesn't happen."

"We're creating alert fatigue across the organization," Gary continued. "Application teams are receiving dozens of recommendations from different sources with no established prioritization, implementation guidance, or streamlined way to address them. Most sit unresolved for months, if they're addressed at all."

"We recently ran a pilot to track what it takes to actually implement a single security control across just ten accounts," Mark explained. "The effort was overwhelming: stakeholder meetings, custom communications, spreadsheet tracking, follow-up emails, escalations, and manual verification. It consumed nearly half my calendar for two weeks."

Gary nodded. "When we extrapolate that effort across hundreds of accounts and the dozens of critical controls we need to implement, the numbers are staggering. We calculated that implementing our priority controls using our current manual approach would require over 8,000 person-hours, equivalent to four full-time employees doing nothing else for a year."

"I see the resource challenge," Charles said thoughtfully. "Before we jump into a new platform, should we just accelerate adding more team members per our longer term plan? I want to make sure we're not adding another tool when this might be a process or organizational issue."

"More people would help," Gary acknowledged, "but it wouldn't solve the fundamental problem. This isn't just about our team's capacity; it's about the overall governance process. Even with double our current headcount, we'd still be using highly manual processes that don't scale with our cloud growth, nor with extra responsibility on the application teams."

Fran took over the financial analysis. "We did a detailed assessment of the manual overhead across all stakeholders, not just the Cloud Team, but also security, finance, operations, and application teams. Using our standard headcount cost model, we're collectively spending nearly $850,000 annually just on governance administration."

"And that's just operational overhead," Gary added. "It doesn't account for the opportunity cost of delayed implementation. Each recommendation that sits unaddressed represents either continued risk exposure or unrealized cost savings."

"So what's the actual business case for this platform?" Charles asked.

"We've identified four primary value drivers," Fran explained, displaying their ROI calculations.

"First, efficiency gains. By automating the governance workflows to handle communications, tracking of status, handling exceptions, etc., we can reduce the manual overhead by at least 30%, representing about $255,000 in annual productivity improvements."

"Second, faster risk mitigation. Currently, it takes an average of 47 days to remediate non-critical findings. With automated workflows and simplified remediation guidance, we can significantly reduce this time, representing a risk and ops time reduction value of approximately $320,000 annually based on our risk management model."

"Third, improved cost optimization. We've already identified $1.2 million in potential annual savings a few months ago, but we've only implemented about 40% of those recommendations due to our manual processes. Increasing implementation rates to 80% would yield an additional $480,000 in annual savings."

"Fourth, reduced organizational friction. The current process creates substantial burden across all teams, with dozens of emails, meetings, and follow-ups for each finding. This operational drag costs us at least $420,000 annually in lost productivity."

"I understand these numbers can be fabricated in multiple different ways," Gary interjected, "But regardless of the specific cost avoidance and savings we calculate, we know we have wasted cloud spend we are tracking and not taking action on that far exceeds the cost of a platform. We have risky misconfigurations we know about that have not been closed, causing our risk profile to incrementally increase. We have a long way to go to bring Goliath to a cloud-first organization where that demand will only grow, and we need to increase our controls further to satisfy Audit. This is why we are trying to scale our Cloud Governance processes to 'Make Change Happen'. A governance platform will help us accelerate this maturity and provide a foundation to make it sustainable for the company."

"I agree with you on why this matters, and can see why we need to automate in order to scale our goals. We just haven't put anything officially in the budget for this, nor has it been in consideration to add yet another platform," Charles stated. "But, what is the alternative? We grow further in the cloud, introduce wasteful spending, increase risk, and burn out resources along the way?" Charles paused, sat back in his chair, and said aloud to himself, "How are we going to fund it with no capital allocated?"

Gary was quick to chime in, "What's interesting about cloud governance platforms is they favor a consumption-based pricing model, unlike security or finance platforms that want to ingest all your data upfront and charge platform fees accordingly. With this approach, we only pay for the controls we're actually using based on the resources they're managing."

He continued, "Cloud governance is progressive by nature. You're constantly iterating on your posture over time, and the cost grows only as you adopt more controls and derive more value. We can start small with just a few key controls in our sandbox environments, prove the value, and scale from there. There's no large upfront commitment."

"As for funding," Gary continued, "while this benefits multiple Shared Services and application teams, having it split across budgets would create unnecessary complexity, nor would Security or other teams pay for a tool that seems to be overlapping with their investment. Since this is core to our cloud governance charter, we recommend funding it from the Operations Shared Services budget. We can allocate the costs through AWS Marketplace and treat the spend as part of our other shared service cloud operating budget."

Charles considered the proposal. "The consumption model reduces our risk, and the business case is compelling. I don't know how we would scale without investing. Let's approve a pilot focused on implementing your next security or cost optimization. Demonstrate measurable value, and we'll discuss expanding from there."

"That works perfectly with our approach," Gary replied. "We'll start small, prove the value, and scale as we demonstrate results."

Day 399: Savings Mandate

"Fifteen percent reduction in cloud spend across all divisions. By the end of next quarter." Fran said in the doorway, startling Gary in his office.

An email from Charles forwarded a mandate from the CIO about a company-wide cost-cutting initiative. Shared Services was tasked with leading the effort, and cloud spending was squarely in the crosshairs as one of the most controllable expenses.

"Travel expenses first, then technology investments," Fran remarked, recalling the message. "Hope this doesn't affect our plans to bring a crowd to CloudNexis."

They'd just started their platform pilot after Charles's approval. Their sandbox environments were connected, and Peter and Dennis had agreed to hook up development accounts. They'd planned to implement the 'No Public AWS EBS Snapshots' control first, but this cost initiative changed the priority.

"Let's pivot to cost optimization," Gary decided. "One control won't satisfy the entire 15% target, but it can showcase the platform's value."

"Which control?" Fran asked.

Gary pulled up their list of cost controls they'd been wanting to implement but hadn't yet deployed at scale. They needed to identify a control that would add up across accounts, wasn't complicated to implement, wasn't complicated for teams to understand, and would offer real savings on wasted spend.

"I'm thinking it should be aging snapshots and backups," Gary suggested. "Teams create them and forget them. The waste accumulates silently. And this is a resource type we had focused on earlier."

"That's perfect," Fran agreed. "Not only would it provide immediate cost savings, but rotating backups and snapshots is a recurring control that auto-remediation can handle. It's an easier progression to having a control continuously run."

"And it's a win for Audit," Gary added. "We can tie it to that IAPP control about backup retention; I believe it was for 90 days. This would be easy to implement, and over time, we can add nuances for shorter time periods for development, or longer durations per workload and environment."

They'd already identified this opportunity through their FinOps tool. The governance platform would add the structured workflow to implement it systematically.

Later that day, Gary and Mark connected their sandbox accounts to the platform to test the control. They set up the control in the "Draft Phase," setting the 90-day threshold, tweaking communication templates, and defining remediation rules.

"Let's walk through the full rollout workflow," Gary said. "Draft, Preview, Check, and Enforce."

They progressed through each phase, sending notices to themselves, receiving alerts, trying self-service actions, and enabling auto-enforcement. The platform identified over 200 snapshots in their small environments, some dating back over a year. Everything worked smoothly.

They decided to let the system run automatically overnight to see how it would perform without their intervention.

The next morning, Gary found the platform had successfully removed another batch of snapshots. The auto-enforcement had worked perfectly.

"Time to try it with Peter and Dennis," Gary decided, messaging them both.

Peter replied immediately: "Perfect timing. We've been talking about cleanup. 2pm work?"

Dennis added: "Curious to see this. Will join."

During the afternoon session, Gary enabled preview mode for their development accounts. Within minutes, the platform identified over 500 snapshots older than 90 days, with an estimated annual savings of $12,000.

"Some of these are from two years ago," Dennis admitted, surprised.

"We've meant to clean these up," Peter added. "But it always falls to the bottom of the priority list."

Gary progressed to the alert phase, sending notifications with self-service remediation options.

"This is much more actionable than our usual alerts," Dennis noted. "Though handling these one by one would get tedious."

"That's why we have auto-enforcement," Gary replied. "For routine maintenance, it's more efficient."

Peter agreed. "After seeing the preview data, I'm comfortable moving straight to enforcement. These are just development accounts, and there's no reason to keep two-year-old snapshots."

With their approval, Gary enabled enforcement mode.

"Let's expand this to RA and EDS next," Mark suggested. "They're already closely aligned with us."

Gary nodded. "Perfect testing ground before we present to Charles. A simple control with proven value and no disruption."

Day 406: Growing Momentum

"In just one week, we've removed nearly 1,100 aging snapshots across four accounts," Gary explained, displaying the dashboard on the conference room screen. "The annualized savings is approximately $38,000, and that's just from a single control in a handful of accounts."

Charles reviewed the metrics carefully. A week after implementing their first automated cost control through the governance platform, Gary had called this meeting to share the initial results. Fran sat nearby, adding context to the financial data.

"Beyond the direct savings," Fran added, "we've established continuous compliance with IAPP control DM-42 for backup retention. The platform is now enforcing the 90-day policy automatically, preventing future accumulation of unnecessary backups."

"I'm impressed with how quickly you implemented this," Charles acknowledged. "Most cost initiatives take weeks just to identify opportunities, let alone implement them."

"That's the key differentiator," Gary explained. "Our FinOps tools have always been good at finding these issues, but the governance platform bridges the gap to actual implementation. We're not just identifying waste; we're systematically eliminating it."

"What feedback have you received from Peter and Dennis?" Charles asked.

"Overwhelmingly positive," Gary replied. "They appreciated the structured approach with transparent communication, easy exception handling, and automated enforcement. Dennis mentioned it was 'refreshingly non-disruptive' compared to previous central initiatives."

"Any resistance to auto-enforcement?" Charles pressed.

"Quite the opposite," Gary said. "After seeing the preview data, they actually preferred to move directly to automation rather than dealing with individual alerts. The preview phase built enough confidence that the enforcement wouldn't cause problems."

Charles nodded thoughtfully. "This aligns perfectly with our cost reduction directive. How quickly could you expand this control to other accounts?"

"We've already begun with Risk Analytics and Enterprise Data Services," Fran reported. "They joined the pilot last week. Since they're newer environments, the savings are more modest. About $12,000 annually from their development accounts."

"Any interest from other divisions?" Charles asked.

"Yea, Jennifer reached out through the Cloud Champions channel." Gary replied. "And I also heard directly from Barry after Dennis spoke with him. Barry is feeling the cost pressures and wants to see what else we can do to help reach his cost savings targets."

Charles nodded. "That's promising. Barry wouldn't invest time unless he saw potential value. OK, what's next?"

Gary pulled up their expansion plan. "We're proposing a three-phase rollout. First, expand to all development and test environments of teams already working with us. Then to non-critical production systems of those same teams. Finally, offer it to other divisions as they express interest."

"How would you quantify the total potential savings?"

"Based on our preliminary analysis," Fran replied, "this single control could yield approximately $180,000 in annualized savings across all accounts. And that's just one of several cost optimization controls we've identified in our earlier FinOps assessment."

"More importantly," Gary added, "the platform approach means we can scale governance without scaling overhead. We can implement multiple controls with the same team, creating a multiplier effect on our impact."

Charles considered this. "You know, when you first proposed this platform, I was concerned it would just turn into another monitoring tool. But this demonstrates a fundamental shift from finding issues to actually solving them."

"That's exactly our goal," Gary confirmed. "Each control we implement becomes a permanent part of our governance baseline, continuously enforcing our policies and preventing drift."

"The timing with the cost initiative couldn't be better," Charles noted. "This gives us a concrete example of how the Cloud Governance Team adds value to the organization."

"And importantly, we're enabling application teams," Gary emphasized. "The platform reduces their burden by automating routine maintenance tasks that they never have time to prioritize."

Charles made a decision. "I want you to accelerate this rollout. Leverage the cost initiative to drive adoption across all divisions. Show concrete savings we can report back to the CIO."

He paused, then added, "And find another high-impact control to implement next. Let's demonstrate that this wasn't a one-off success but a scalable approach to governance."

Day 413: Barry's Pivot

Barry reviewed the latest implementation metrics from the snapshot cleanup control. In just one week, Commercial Banking had removed 843 aging snapshots, yielding $22,000 in annualized savings with virtually no disruption to their development teams.

"The automated enforcement is working well," Mike reported during their morning update. "No manual intervention or exceptions required from our side."

Barry nodded, tapping his pen against the polished surface of his desk, a habit that over his twenty years at Goliath had become a tell that he was genuinely pleased with something. "What's next?"

"Gary's team is proposing idle resource shutdown as their next control," Mike explained. "They've estimated potential savings of $180,000 annually for our cloud environments."

Barry considered this thoughtfully. The cost reduction mandate from the CIO was putting unprecedented pressure on his budget. Commercial Banking had aggressive delivery goals that couldn't be compromised, yet they needed to find substantial savings.

"Let's proceed with it for development environments," Barry decided, his tone matter-of-fact where once there would have been hesitation. His voice lowered, "But let me be absolutely clear, this stays far away from production. No reason to shut down any resources there at the moment."

Mike looked slightly surprised. "That's more receptive than I expected."

Barry glanced at the award shelf behind his desk. It was a collection of achievements representing Commercial Banking's history of successful delivery under his leadership. "The cost pressure leaves us little choice," he replied pragmatically. "We need to find savings that don't impact our delivery capacity. This approach is working." He paused. "And frankly, after years of dealing with fragmented requests from different central teams, having one consolidated approach is refreshingly efficient."

That was the real difference. Before the governance platform, Commercial Banking had been bombarded with disjointed requirements from Security, Finance, Operations, and Audit, each with their own tools and priorities. Now they had a single channel that was focused and consistent.

"The team appreciates the streamlined process," Mike confirmed. "Rather than juggling multiple different compliance requirements, we're getting coordinated guidance."

"That's why this is working," Barry observed, leaning forward. "Gary's team is functioning as a buffer between us and all the other noise, while still ensuring we maintain compliance."

As Mike left, Gary stayed behind with a request. "The platform metrics show impressive results for Commercial Banking. I'm preparing our CloudNexis presentation in a few months. I think it would be great to include our progress highlighting your division's achievements."

Barry leaned back in his chair, the same chair from which he had rejected countless central IT initiatives over the years. He reflected on Commercial Banking's position, their proud history of autonomy now evolving into something he wouldn't have predicted being showcased publicly as leading Goliath's cloud journey.

"Sure, it would be great to feature the work Commercial Banking has achieved," he replied, surprising even himself with his ready agreement. "Send me the slides for review before finalizing."

A small smile crossed his face as Gary left. Barry had always measured initiatives by how they affected his team's ability to deliver business value. Now he was realizing how his growing support had enabled much more value for the organization working together versus in silos. The question of who controlled what, once so important, now mattered less than what they accomplished together.

Day 420: Platform Commitment

"Two weeks, two controls, and tangible results," Gary summarized, presenting the dashboard metrics to Charles. After their initial pilot with the snapshot cleanup control and the recent expansion to idle resource management, this meeting would determine their formal commitment to the governance platform.

"The numbers are compelling," Charles agreed, studying the data. "Nearly $100,000 in annualized savings identified already, with minimal disruption to teams."

Gary nodded, feeling confident in their approach. "We strategically chose cost optimization controls for our initial pilots because they demonstrate immediate, tangible ROI that's easy to measure. The snapshot cleanup and idle resource management directly translate to dollar savings that help justify the platform investment upfront."

"Smart approach," Charles acknowledged. "Start with controls that pay for themselves immediately."

Charles addressed Fran and Sarah directly. "Fran, Sarah, you both have significant investments in your FinOps and CNAPP platforms. Do you still feel this governance platform complements rather than competes with your existing tools?"

"Our FinOps platform is excellent at identifying cost optimization opportunities, but we have no systematic way to assign these to teams, track implementation, or measure outcomes," Fran replied. "The governance platform bridges that gap. It takes our FinOps recommendations and turns them into actionable workflows."

"Similar story with our CNAPP tool," Sarah added. "We have broad security scanning visibility to identify recommendations, but this governance platform is helping us make them actionable and work with the application teams to close them out."

Charles nodded thoughtfully. "So you're not replacing these tools, you're making them more effective."

"That's right," Gary replied. "Each control has its own value proposition. The cost optimization controls show direct savings that offset the platform subscription costs. For instance, the snapshot cleanup control costs approximately $3,000 annually in platform fees but yields nearly $87,000 in savings, nearly a 30:1 return!"

Gary turned to the next phase of their strategy. "Now that we've validated the approach with these initial controls, we want to focus on migrating our existing security controls to the platform. While these may not show the same direct cost savings, their value is equally substantial when properly quantified."

Sarah jumped in, "The security controls we've been trying to implement manually for months represent significant risk reduction. For example, our CNAPP tool has identified over 300 security misconfigurations that we simply haven't had the bandwidth to address systematically."

"And there's a real cost to those unaddressed issues," Gary emphasized. "When we applied the risk valuation model based on industry data, just the exposure risk from unaddressed security findings represents approximately $1.2 million in potential incident response, remediation, and regulatory penalties."

"But beyond the risk reduction," Sarah continued, "there's also the operational efficiency component. Each security control we implement manually consumes substantial resources from both the Cloud Team and application teams. We're spending an average of 40 person-hours per control implementation across just ten accounts."

"The platform reduces that to about a few hours of configuration time," Gary added. "That's an 87% reduction in overhead, freeing up resources to focus on higher-value activities instead of repetitive governance tasks."

"And importantly," Sarah emphasized, "these are controls we've already determined we need based on our audit requirements and risk assessments. Upfront, we're not adding new compliance burdens; we are simply implementing existing controls more efficiently and consistently."

Mark joined in, "We've developed a standardized process to evaluate each control. For cost optimization controls, we calculate direct savings against platform costs. For security controls, we quantify risk reduction value and operational efficiency gains. And for operational controls, we measure improvements in reliability and performance."

"What I'm hearing," Charles summarized, "is that you're approaching governance controls as individual business decisions, each with its own value proposition, rather than a massive all-or-nothing investment."

"That's precisely it," Gary confirmed. "By evaluating each control on its own merits, we can prioritize those with the highest impact while maintaining the flexibility to adjust our approach as needs evolve."

Fran added the financial perspective. "The consumption-based pricing model aligns perfectly with this approach. We pay only for the controls we implement and the resources they manage, so costs scale directly with value delivered."

Charles considered this for a moment. "If we're proceeding with this model, we need a structured review process to ensure we're realizing the expected value and making informed decisions about which controls to implement next."

"We're proposing monthly check-ins to review active controls and their impact," Gary replied, "along with a pipeline of upcoming controls we're evaluating for implementation."

"And quarterly reviews of the overall platform costs," Fran added. "As our usage scales, we should evaluate opportunities for volume discounts through longer-term commitments. We can align this with our broader FinOps process for cloud cost optimization."

Charles nodded, satisfied with the approach. "That's sensible. Start with the consumption model to prove value, then consider more committed pricing once we understand our steady-state usage patterns."

"I think we have alignment then," Charles concluded. "Proceed with converting to a paid subscription, focusing initially on migrating those existing security controls that you've already identified as critical but haven't been able to implement effectively. We'll review progress monthly and assess the economic model quarterly as we scale."

Day 431: Governance Review, Part One

"Let's take stock of where we are," Gary announced, bringing up a dashboard on the conference room screen. Eleven days had passed since their formal commitment to the governance platform, and today would be their first structured governance review, now establishing the process they'd promised Charles they would follow.

The core team had assembled for the review session, with representatives from Security, Finance, Operations, Asset Management, Networking, and Identity Management.

"I've organized this review into four sections," Gary explained. "Account onboarding status, active controls evaluation, application team feedback, and future control planning."

Mark began with the account integration status. "Out of our 423 mapped AWS accounts, we've successfully connected 287 to the platform. For Azure, we've connected 42 out of 68 subscriptions, and for GCP, 16 out of 23 projects."

"What challenges are we seeing with the remaining accounts?" Gary asked.

"Three main issues," Mark replied. "Legacy accounts with non-standard configurations, specialized environments with unique permission requirements, and accounts where we're still coordinating with application teams."

Isaac nodded. "Some of those accounts we haven't fully brought into our governance process as we're still dealing with pushback from certain business units, though we are making progress. Others were created completely outside our process that we've caught. One item I want to bring up, Gary, is that we need to create a workstream to communicate more proactively about the cloud program to help educate and train folks on cloud and the process we have."

"That's a good point," Gary replied. "Let's park that discussion for now and circle back soon on how to tackle that. It's definitely something we need to address."

Gary turned to the control implementation progress. "Let's look at how our active controls are performing."

Sarah displayed their control migration dashboard. "We've successfully implemented the two cost optimization controls from our initial pilots, snapshot cleanup, and idle resource management. We've also migrated the 'No Public AWS EBS Snapshots' control that we've been piloting. Currently, we're in the process of migrating seven more controls from our existing framework that align with CIS security recommendations and map to our IAPPs."

"The cost controls are active across 47 accounts, primarily development and test environments," she continued. "The 'No Public AWS EBS Snapshots' control is deployed to 28 accounts so far."

"That seems limited compared to our total account footprint," Oscar observed.

"We're being intentionally methodical," Gary explained. "Following our rollout approach, we're moving controls through phases across different environments, starting with development accounts."

"How are the controls performing?" Fran asked.

"The snapshot cleanup control has processed over 2,400 resources in total, yielding approximately $110,000 in annualized savings, tracking ahead of our initial estimates."

"For the security controls, we're seeing a 92% reduction in remediation cycle time compared to our manual processes," Sarah added. "Controls that previously took weeks to implement are now being deployed and enforced in hours."

Gary turned to feedback from various application teams. "We've been collecting their input through the Cloud Champion's Microsoft Teams channel and one-on-one conversations."

Mark displayed a summary. "Overall, the response has been positive. Teams particularly appreciate the self-service remediation options and transparent communication."

"Several teams have reported frequent security group modification requests to allow access to port 53 for internal domain name DNS resolution," he noted. "They've suggested this could be included in our default security group setup in the landing zone."

"Instead of handling these requests individually, we could update our landing zone network templates," Nancy replied. "This would reduce friction and standardize configurations."

Adam nodded. "And we could potentially migrate some of these landing zone configurations to the governance platform as policies, allowing us to communicate changes more effectively and automatically remediate drift."

Isaac added, "It would also help maintain consistent identity and access patterns. Many exceptions relate to standard service account access that could be templated."

Gary made a note. "This is exactly the kind of insight I was hoping would emerge, using what we learn to improve our foundation. I have taken a note for us to work on the next landing zone improvement, and for Isaac and I to discuss an overall communication plan to get in front of more divisions on cloud at Goliath."

Gary looked around the room, noting the engaged faces and productive discussion. The review was proving valuable; not just for tracking progress, but for identifying opportunities they hadn't considered.

"Let's shift gears and talk about future controls," he announced. "We need to evaluate what comes next after we complete our current migration efforts."

Day 431: Governance Review, Part Two

Gary pulled up their prioritized list of potential controls, shifting the discussion from current progress to the future direction.

"We need to evaluate what comes next after we complete our current migration efforts," he began. "Let's assume we could implement just two new controls per month to allow time for teams to adapt while maintaining steady progress."

Fran displayed their analysis, extracted from monitoring tools. "We've aggregated recommendations from our FinOps platform, CNAPP tool, and operational monitoring systems."

"We've tried developing scoring models, but honestly, the prioritization comes down to judgment calls," Gary admitted. "Do we focus on the $400K in cost savings from rightsizing, the critical security gaps, or the foundational controls that enable everything else?"

Sarah pushed back immediately. "We have over 300 unaddressed security findings from our CNAPP tool. How is cost optimization a higher priority than critical security gaps?"

"Because the business is under pressure to reduce spending," Fran countered. "Cost controls show immediate, measurable value that we can report to Charles and the CIO."

"But security risks compound over time," Sarah replied. "Every day those misconfigurations remain unfixed increases our exposure."

Oscar raised another concern. "Network controls don't show direct financial impact, but they're foundational for everything else. How do we account for controls that enable other controls?"

"There's no simple formula for this," Gary acknowledged, looking at their potential controls list. "We could prioritize by financial impact, security risk, regulatory requirements, or organizational readiness. Each approach yields different answers."

"So how do we actually decide?" Nancy asked.

"Judgment calls," Gary replied. "Informed by data, but ultimately based on what we think the organization can absorb and what delivers the most value right now."

Oscar raised a practical concern. "Now that we have the platform, why limit ourselves to two controls per month? We could enable a dozen new controls tomorrow and catch up faster."

"That's exactly the trap we're trying to avoid," Gary replied. "Six months ago, we would have turned on every available control and pushed all the alerts to the application teams."

"But we learned that teams can only absorb so much change," Sarah added. "Even with perfect communication and self-service remediation, there's a limit to how many new requirements they can handle."

"It feels slower," Mark admitted, "but teams are actually fixing things instead of ignoring them."

Gary paused, recognizing the shift in their thinking. "This is fundamentally different from how we used to work. Before, we'd identify problems and immediately push them to teams. Now we're deliberately choosing what not to implement."

"We're controlling the rollout," Mark observed. "Instead of flooding teams with alerts, we're being strategic about change management."

"The challenge is that being strategic feels like being slow," Fran noted. "When leadership asks why we haven't implemented more controls, it's hard to explain that restraint is actually progress."

"But that's exactly what governance maturity looks like," Gary said. "Moving from reactive to strategic. From identifying everything wrong to systematically improving what matters most."

The conversation turned to their immediate next steps after completing the security control migration.

"What about the AI model governance controls that Robert from Risk Analytics requested?" Sarah asked. "They don't affect as many accounts, but they're critical for those specific business initiatives."

"That's an important point," Gary replied. "Our prioritization helps with broadly applicable controls, but we also need to support emerging technologies. The AI governance controls may not have the widest impact today, but they're strategically important."

"We should consider creating a separate track for these specialized controls," Mark suggested.

"I like where we are going with this discussion," Gary said, "but let's park this for our next session so we can stay focused on completing the migration. There's plenty of time to discuss prioritization strategy in our next review meeting."

As they wrapped up their first structured governance review, Gary reflected on how well the session had gone. They'd moved from reviewing progress and discussing user feedback to grappling with strategic questions about change management and organizational capacity. They were laying a solid foundation for continuous improvement cycles that would strengthen their cloud governance going forward.

Day 435: Change Absorption Capacity

"Why are we limiting ourselves to two controls per month?"

Oscar's question from the governance review meeting had been nagging at Gary for days. It seemed like a reasonable challenge. They had the platform, they had identified dozens of needed controls, so why move so slowly?

Gary had called Sarah and Mark into his office to hash this out properly.

"I keep coming back to Oscar's point," Gary began. "We have the capability to deploy controls faster. Are we being artificially conservative?"

Sarah pulled up her laptop, showing their current deployment timeline. "Let's think about what 'two controls per month' actually means for teams. Each control requires them to understand new requirements, potentially modify their processes, and adapt their workflows."

"Even with our rollout approach, each phase requires mental bandwidth," she continued. "'Preview Phase' means they need to review and understand the impact. Check phase means they're getting alerts to triage. Enforce phase means automatically implementing changes."

Gary gestured to the timeline. "It's not just about technical complexity; it's about organizational overhead of processing change."

"Exactly. And that overhead varies by team maturity," Sarah agreed. "Peter's team can probably handle more frequent changes because they have established processes. A team new to cloud might struggle with even simple controls."

Mark nodded. "We learned this with Dennis's team early on. When we started piloting multiple controls simultaneously, they got overwhelmed trying to keep track of everything."

"There's another constraint we haven't discussed," Gary noted. "Our own team capacity. We can only manage so many controls in different rollout phases simultaneously."

Mark pulled up their current tracking board. "Right now, we're managing 'Preview', 'Check', and 'Enforce' phases across multiple controls. Each phase requires different types of support from us."

"And controls won't progress evenly across accounts anyway," Sarah added. "Newer accounts might skip straight to enforcement if they don't have existing violations. Mature accounts like Peter's might advance faster through the phases."

Gary moved to the whiteboard. "We're managing multiple rate limits: team absorption capacity, our operational capacity, and the complexity of having different accounts at different phases."

"What signals tell us a team can handle more change?" Mark asked.

Gary thought about their recent experiences. "Remediation speed is one indicator. If teams consistently fix issues quickly, they probably have bandwidth for more. Exception requests could be another where teams frequently requesting exceptions might be at their capacity."

"And proactive engagement," Sarah suggested. "Teams that actively engage and ask for additional controls are ready for more change."

Gary began sketching on the whiteboard. "So we start conservative, then scale based on team maturity, control complexity, and capacity indicators."

"We could categorize controls by impact level too," Mark suggested. "Simple cost optimizations could deploy more frequently than complex security policies requiring process changes."

"And we could have different cadences for different teams," Sarah added. "Mature teams might handle monthly changes, while newer teams need quarterly updates."

Gary stepped back from the whiteboard. "The principle becomes: Start with a conservative pace, then scale based on demonstrated success, both theirs and ours."

"This gives us a framework for explaining our approach," Mark noted. "When teams ask why we're being 'slow,' we can explain we're optimizing for sustainable adoption rather than just speed."

Gary closed his notebook, satisfied with their framework. The question wasn't whether they could deploy controls faster; it was whether teams could successfully absorb and implement those controls while maintaining their primary responsibilities.

Day 437: Preventive vs. Detective Controls

"Should we prevent teams from creating public S3 buckets, or just detect and fix them afterward?" Sarah asked, pulling up a specific example on her laptop.

Two days after their change absorption discussion, Gary had gathered the core team to tackle another fundamental governance question: how much should they prevent versus detect and correct?

"It's the classic security versus agility debate," Gary replied, studying the example. "But I think we're approaching this wrong. It's not just about risk level; it's about what happens when we get the decision wrong."

Mark looked puzzled. "What do you mean?"

"Prevention can break legitimate automation," Gary explained. "What if a deployment pipeline or auto-scaling group automatically creates resources that don't initially meet our exact specifications? Prevention could fail the entire process."

Sarah nodded slowly. "That's exactly what happened with the Risk Analytics team. Their ML training jobs automatically create temporary S3 buckets that start public for data ingestion, then get locked down. A prevention control would have broken their entire workflow."

"The question isn't just 'is this risky?'" Gary continued, moving to the whiteboard. "It's 'can we safely and quickly fix this after creation, and what breaks if we prevent it?'"

"But some things genuinely should be prevented," Sarah countered. "Like deploying resources in an unapproved region. You can't move resources to another region, so the only remediation is to delete and recreate."

"That's the key distinction," Gary said, writing on the whiteboard. "Prevention makes sense when remediation is difficult, disruptive, or impossible. Detection works when we can correct issues quickly without major impact."

Gary began sketching examples. "Prevention scenarios would be global constraints like blocked services or prohibited regions, and configurations that can't be changed after creation or require deletion and recreation to fix."

"Detection scenarios," Sarah continued, "would be settings that can be corrected in-place, configurations where automation might legitimately create non-compliant resources initially, and policies where 'correct' depends on context that develops over time."

"Right, plus situations like budget overruns, resource quotas, and monitoring utilization," Mark added. "You can't prevent those - they have to build up before you can act on them."

"There's another factor," Sarah noted. "How quickly can we detect and remediate? If we can auto-fix a public S3 bucket within minutes, the risk window is minimal."

"That changes the equation completely," Gary agreed. "A high-risk configuration that gets auto-corrected in thirty seconds might be safer than a medium-risk configuration that requires manual intervention."

Mark was nodding. "It's really about the end-to-end risk profile, not just the initial risk level."

"Think of it like traffic management," Gary suggested, updating his whiteboard analogy. "Traditional governance is like gates; you have to stop and get permission before proceeding. That's sometimes necessary, but it kills velocity."

Sarah continued the analogy. "Guardrails let you drive fast safely. They guide you away from dangerous areas but don't force you to stop unless you're heading for real trouble."

"And detective controls with auto-remediation are like modern safety systems," Mark added. "Lane departure warnings, automatic braking. They let you operate freely but intervene when needed."

Gary stepped back from the whiteboard. "Then our approach becomes: preventive controls for global constraints and when fixing later is difficult, everything else gets handled by auto-remediation guardrails to mitigate risk while maintaining agility."

"And be explicit about remediation speed," Sarah suggested. "If we can't auto-fix something within a reasonable timeframe, maybe it should be prevented after all."

"This also helps with change management," Mark noted. "Teams are more likely to accept detective controls because they preserve operational flexibility while still managing risk."

Gary reviewed their framework. "The key insight is that we're not choosing between security and agility. We're optimizing the risk-agility tradeoff by being smart about when and how we intervene."

"And recognizing that modern cloud governance can be more sophisticated than traditional gate-based approaches," Sarah added.

"Most things can be handled through guardrails," Gary concluded. "Prevention should be the exception, not the rule."

Day 441: Remediation Challenge

"What do you mean the test database is offline?" Gary asked, the urgency in Dennis's voice immediately putting him on alert. It was 7:30 AM, and he'd barely settled at his desk when the call came in.

"The security group rule standardization control suddenly started enforcing in our dev environment," Dennis explained. "It removed a rule that our database cluster depends on. The developers can't access their data."

Gary felt his stomach tighten. Ten days after their governance review, they'd been making steady progress migrating controls to the platform. The control had passed through the "Preview" and "Check" phases, but enforcement wasn't supposed to affect development environments this way.

"How did the enforcement behavior change?" Gary asked, already pulling up the dashboard.

"That's what we're trying to figure out," Dennis replied. "The account was tagged as 'environment:developmet' yesterday, and our control logic applies different standards to development environments."

Gary quickly messaged Sarah and Mark to join an emergency bridge call. Within minutes, they were piecing together what had happened.

Sarah checked the audit logs. "Found it. A developer updated the environment tag from 'development' to 'developmet' at 6:42 PM yesterday, likely a typo. But our control logic only looked for exact matches, so it didn't recognize this as a development environment."

"The good news is this only affected a development database," Dennis said. "If this had happened in production, it could have been much worse."

Mark temporarily muted the enforcement for Dennis's account. By 9:00 AM, the database access was restored. Gary called a debrief session with the broader team to understand what happened and how to prevent similar issues.

“Our control logic was too rigid. We designed it to look for specific tag values without considering common variations or human error," Mark explained.

"This highlights a fundamental challenge with automated remediation," Sarah observed. "The platform executed exactly as configured, but our control logic didn't account for the messy reality of how teams actually use tags."

"What's interesting," Dennis noted, "is how much easier it was to identify and fix this issue compared to our old approach. The platform captured exactly what happened, when, and why the remediation triggered. Before, we would have spent hours just trying to figure out what went wrong."

Oscar nodded. "But there's a deeper design question here. Should we be making policy decisions based on user-controlled metadata at all? Tags are meant to be descriptive, not deterministic for security controls."

"I think we need to fundamentally rethink our approach," Mark suggested. "We could make the logic more flexible by recognizing variations like 'dev', 'development', 'developmet' and 'sandbox' as equivalent. But the bigger issue is whether we should be using changeable metadata to make critical security decisions at all."

"We could do more extensive testing in 'Preview'," Sarah suggested. "But realistically, we'll never catch every edge case. The key is building controls that fail safely and can be quickly identified and corrected when issues arise."

"That's exactly right," Gary agreed. "This incident shows we need more resilient control logic, not just better testing. We should assume human error will happen and design accordingly."

Oscar raised an important point. "This incident highlights why we need to separate the concept of 'environment classification' from 'policy enforcement.' Teams should be able to correctly label their resources without accidentally changing their security posture."

"That's a critical insight," Gary acknowledged. "We need to distinguish between descriptive metadata controlled by teams and policy assignments enforced by governance processes."

As Gary updated Charles on the incident and their response, he emphasized the broader lesson: automated governance required more sophisticated thinking about policy logic and failure modes than manual processes.

"What impressed me most," Charles noted after reviewing the incident report, "was how the team responded. Instead of backing away from automation, you're figuring out how to make it more resilient."

Day 443: Policy Logic and Control

"We need to talk about control logic," Gary said, settling into the conference room where Sarah, Mark, and Oscar had gathered for a follow-up discussion.

Two days after Dennis's team had experienced the automated remediation triggered by a tag misspelling, the implications were still reverberating through the Cloud Governance Team's thinking about policy design.

"The incident exposed something fundamental about how we're designing our controls," Mark began, pulling up their current policy configurations. "We're writing logic that assumes perfect conditions without thinking through all the scenarios."

Sarah nodded. "Our security group control was designed to recognize 'development' environments and apply relaxed rules. But we only coded for the positive case - what happens when the tag matches. We didn't think through what happens when it doesn't match."

"It's like we wrote an allowlist without considering the blocklist implications," Gary observed, moving to the whiteboard. "When the tag said 'development', the control knew what to do. When it said 'developmet', it fell back to default behavior, which was production-level enforcement."

Oscar pointed to the whiteboard. "That's a design pattern we need to fix across all our controls. Every conditional policy should explicitly handle both the true and false cases, not just assume one or the other."

"But there's a deeper issue," Mark continued. "Should we be basing security controls on metadata that developers can freely modify? When a tag determines security posture, we're essentially letting application teams dictate their own governance requirements."

Gary began sketching on the whiteboard. "That's the real problem. If security posture is defined by a tag, then changing that tag changes security policy. We've given developers indirect control over their compliance requirements."

"And tags provide infinite scenarios," Sarah added. "Teams can create any value they want. 'dev', 'development', 'sandbox', 'test', 'staging', 'prod-like.' How many variations do we need to account for?"

"We could use the platform's auto-tagging features to standardize this," Mark suggested. "But that still doesn't solve the fundamental question of whether user-controlled metadata should drive policy decisions."

Gary stepped back from the whiteboard. "I think we need to separate two concepts: resource classification and policy assignment. Teams should be able to classify their resources however makes sense for their operations. But policy assignment should be based on more authoritative sources."

"Like account structure," Oscar suggested. "If an account is designated for production workloads through our Account Gateway, it gets production policies regardless of how teams tag individual resources."

"Or explicit policy designations that require approval to change," Sarah added. "Teams can't just modify a tag and alter their security requirements."

"This changes how we think about control design," Mark noted. "Instead of asking 'what tag value triggers this policy,' we should ask 'what authoritative signal indicates which policy applies.'"

Gary turned back to the whiteboard, updating his diagram. "So our principle becomes: base policy decisions on authoritative signals, not on team-managed metadata."

"And when we do use conditional logic," Sarah added, "we need to explicitly define behavior for all possible conditions, not just the ones we expect."

"It's about defensive programming for governance," Oscar observed. "Assume the unexpected will happen and design accordingly."

Gary closed his notebook, feeling like they'd identified a critical improvement to their approach. The tag incident had been disruptive, but it had forced them to think more rigorously about the logic underlying their governance controls.

"This applies to every control we design going forward," he concluded. "Policy assignment should be deliberate and controlled, not dependent on free-form user input."

Day 448: Exception Process

"We need to talk about what happens when controls don't fit," Gary said, sketching on the whiteboard in the conference room. "Not just the edge cases, but the everyday reality of managing alerts and exceptions."

A week after the tag misspelling incident, the Cloud Governance Team had convened to address a growing challenge: how to handle the flood of alerts that didn't require immediate action, shouldn't be ignored indefinitely, but also didn't need policy changes.

"I'm getting the same question from multiple teams," Sarah began. "How long can they leave a security finding open before it gets escalated? And what's the difference between muting an alert versus requesting an exception?"

Gary turned to the whiteboard. "Let's start with the most common scenario: alerts that teams acknowledge but can't fix immediately. How long should those stay in alarm status?"

Mark pulled up their current findings dashboard. "We have over 200 open alerts right now. Some are two days old, others are two months old. Teams don't know what's acceptable, and we don't have consistent guidance."

"That's the SLA question," Gary acknowledged, writing on the whiteboard. "Without defined timeframes, teams either panic about minor issues or ignore critical ones indefinitely."

Sarah nodded. "And different types of findings need different SLAs. A public S3 bucket with customer data should be fixed within hours. A missing tag in development can wait weeks."

"But who sets those SLAs?" Oscar asked, joining the discussion. "And what happens when they're exceeded?"

Gary considered this. "We need risk-based SLAs that account for both technical severity and business context. Critical findings in production might get 24-hour SLAs. The same finding in development might take 30 days."

"And established escalation paths," Sarah added. "When SLAs are missed, what happens? Auto-remediation? Management notification? Account restrictions?"

"There's another layer to this," Mark noted. "Sometimes teams want to acknowledge an alert and suppress it temporarily. They're aware of the issue, but need more time to address it properly."

Gary wrote "Alert Muting" on the whiteboard. "That's different from ignoring the alert. They're taking ownership but asking for breathing room by suppressing the notifications."

"Exactly," Sarah agreed. "It's like saying 'I see this, I'm planning to fix it, but I need two weeks to coordinate with other systems.' They want to extend their SLA without the alert noise."

"But even muting needs limits," Gary continued. "You can't suppress alerts indefinitely. How long is reasonable?"

"Depends on the finding type and complexity," Mark suggested. "Simple configuration changes might get a few days. Complex architectural changes might need weeks."

Oscar raised another scenario. "What about situations where teams don't want the alert at all? Like Dennis's disaster recovery snapshots that are intentionally shared. That's not about timing; that's about policy."

Gary moved to a different section of the whiteboard, writing "Policy Exceptions." "That's fundamentally different. Muting says 'this is a problem I'll fix later.' Exceptions say 'this isn't a problem for my specific use case.'"

"Policy exceptions change what compliance means," Sarah explained. "Instead of getting alerted about shared snapshots, Dennis's account would be compliant when snapshots are shared with approved partners."

"So we have three scenarios," Gary summarized, pointing to his whiteboard sections. "Alerts within the SLA that teams are actively working on. Alerts beyond the SLA that need muting to extend response time. And legitimate business cases that need policy exceptions to avoid false alerts entirely."

"What about auto-remediation?" Mark asked. "Where does that fit in this framework?"

"Auto-remediation could be triggered by missed SLAs," Gary replied, adding it to the whiteboard. "If a critical finding isn't addressed within the SLA window, the system could automatically fix it rather than continuing to escalate."

"That makes sense for safe-to-fix issues," Sarah agreed. "But teams should be able to opt into that behavior rather than having it forced on them."

"The approval processes should match the scenario," Mark suggested. "Active alerts just need tracking. Muting requests need justification and time limits. Policy exceptions need business and security review."

"And different urgency levels," Sarah added. "Critical findings might auto-escalate after missed SLAs, or even trigger automatic remediation. Lower-priority findings might just generate reports to management."

Gary stepped back from the whiteboard, reviewing their framework. "The key insight is that not every alert requires immediate action, but every alert needs a defined path to resolution within defined timeframes."

"This gives teams predictable options," Oscar noted. "They know what they can handle themselves versus what needs approval, and what the time expectations are for each path."

"And it prevents alert fatigue," Sarah concluded. "Instead of constant noise, teams get actionable notifications with detailed next steps and reasonable timeframes."

Gary felt satisfied with their approach. They'd moved beyond binary compliance to a nuanced system that acknowledged the complexity of real-world operations while maintaining governance oversight.

Day 455: Landing Zone Evolution

The Society of Corporate Compliance and Ethics (SCCE) defines compliance as

Following the Cloud Governance review meeting a few weeks ago, the team had gathered to discuss how to evolve their Account Gateway landing zone approach.

"Even with our standardized Account Gateway process, configurations drift over time," Mark explained. "We set up the right logging configuration initially, but someone later modifies it manually or disables it entirely. We only discover the issue during our quarterly audits, if at all."

Oscar nodded in agreement. "It's happening with security groups and IAM policies as well. What we configure during landing zone deployment isn't necessarily what exists six months later."

This was the fundamental challenge with their current approach. The Account Gateway created a consistent starting point, but it was a one-time event with no ongoing enforcement.

"We've been thinking about landing zones and governance controls as separate things," Gary observed. "Landing zones set up the environment, and then governance controls monitor it. But there's a disconnect."

Sarah had an idea. "What if we reimagined our landing zone patterns as governance controls themselves? Instead of just setting up an environment and handing it off, the control would continuously enforce the required configuration."

Mark pulled up a specific example on his laptop. "Take Account 472 from the Wealth Management team. The landing zone correctly configured CloudTrail with multi-region logging to our central security bucket. Last week, someone disabled it, triggering an alert from our CNAPP tool."

"Exactly," Gary agreed. "If CloudTrail logging was implemented as a governance control rather than just a landing zone component, it would not only configure logging during setup but detect and correct any drift afterward."

"There are resource-level patterns we can't pre-configure during account setup because the resources don't exist yet," Nancy added.

"Like S3 bucket access logging," Isaac agreed. "We can't set that up in advance because buckets come and go. But a control could detect when a new bucket is created and automatically enable access logging."

Gary started sketching on the whiteboard. "I'm seeing three major categories of patterns we could implement as controls: account-level configurations like IAM roles, region-level configurations like CloudTrail, resource-level patterns like S3 bucket logging."

"This would let us move from one-time setup to continuous enforcement," Adam noted. "The landing zone wouldn't just be a starting point; it would become an enforced baseline that prevents drift."

The team spent the next hour identifying landing zone components that could be reimplemented as governance controls. The list was extensive: IAM roles and policies, CloudTrail and CloudWatch configurations, VPC flow logs, security group rules, S3 bucket policies, and many others.

"There's another benefit to this approach," Gary observed. "When we update our landing zone patterns, currently we have no way to apply those changes to existing accounts. But if they're implemented as controls, we can use our rollout process to roll out updates across all environments."

"That's huge," Sarah agreed. "We've been struggling with how to implement new security requirements across existing accounts. This gives us a systematic way to do that."

"And we can use the same tiered approval approach we've established for other controls," Mark added. "Preview phase to see the impact, check phase for validation, and then enforcement."

Gary chimed in, "and even if not all controls apply to all accounts, we could associate specific controls that are attached to specific accounts. Such as an account which needs AI workloads, or another that is approved to deploy public facing workloads. And those controls could be enabled when needed, not just during the initial account setup. If you decide to start using AI six months after an account is created, we can associate the relevant controls at that time."

By the end of the session, they had a clear vision for integrating landing zone patterns into their governance framework. They would start by implementing CloudTrail configuration as a control, then systematically convert other landing zone components into the governance platform.

"This is a fundamental shift in how we think about environment setup," Gary summarized. "We're moving from point-in-time configuration to continuous enforcement of our baseline standards."

"And it closes a critical gap in our governance approach," Sarah noted. "We've been focusing on identifying and remediating issues, but this prevents them from occurring in the first place by maintaining the correct configuration automatically."

"And it gives us a deployment pipeline for infrastructure standards," Mark added. "When security requirements change or we identify better patterns, we can roll them out systematically through the governance framework we've already established."

Day 462: Governance Roadshow

Gary reviewed his notes from the governance meeting held a few weeks ago, focusing on the action item he'd committed to address: developing a proactive communication strategy for cloud services across the organization. Isaac's comment during that meeting had stuck with him. Afterward, Gary followed up to understand how a developer in Claims Processing hadn't even known that cloud services were officially supported at Goliath.

"Morning, Mark," Gary said as his cloud architect entered the conference room. "Thanks for making time to discuss the communication strategy we talked about in the governance review."

"Happy to help," Mark replied, settling into a chair. "I've been thinking about this since Isaac raised it. While we have made good progress across many teams at Goliath, most of the teams still don't know what cloud options are available to them or how to get started. It's almost like we need to 'market' our services to our 'customers'."

"Agreed," Gary nodded. "I had spoken with Charles about this during my last 1:1, and it's something we should start focusing on soon. In fact, it's perfect timing because we have Elena from Enterprise Architecture moving to the team soon to help add capacity given our growing demand.

"That's great news," Mark said. "We've been stretched thin trying to handle the governance platform rollout while also supporting teams that want to migrate to cloud. Having dedicated capacity for enablement and education will make a huge difference."

"Many teams simply don't know what resources are available or how to access them properly. We have had months of building out this offering, it's helpful for teams to be aware of how they can take advantage of it."

"Also, from what we have seen from the RA and EDS teams, many are still new to cloud. We should consider providing general awareness about cloud to address the knowledge gap our other teams have, and couple that with education on how Goliath has extended cloud services with centralized account onboarding, landing zones, and automated governance controls."

Gary stood up to start outlining some concepts on the whiteboard. "I'm thinking these sessions should include several components: an introduction to how cloud services are used at Goliath, success stories from teams already using cloud services, best practices for security, operations and cost optimization, and demonstrations of the Account Gateway and governance platform."

"Let's start with a monthly virtual training session for broad reach," Gary summarized. "Each month gives us the opportunity to refine and add more content as we go. Recording the sessions will help audiences in different time zones to watch later, and eventually we can use them as assets for a library of materials."

"I agree virtual is the best starting point," Mark said. "But, often folks attend virtual events multi-tasking, and in general, it's hard to engage with a virtual audience. I would like to see these sessions be more interactive." Mark paused for a moment.

"Conducting recurring in-person sessions would garner more impact. Perhaps extending the session to be a hands-on workshop to help folks get started. And if we are in-person, we can involve a local cloud champion and other application teams already using the service to share their use cases."

"Yes, great idea!" Gary said enthusiastically. "Getting in front of the users will be a great way to foster those relationships with different teams, especially across our other office locations, where we don't often have the chance to meet each other in person. The workshops will also help us gather feedback on the user experience in real time as users are trying to perform actions in their environment."

The meeting continued, where they outlined additional communication channels to build an intranet site to host a library of materials for their teams to reference, along with a new Teams channel to be used more broadly at Goliath for all things cloud-related.

"When Elena joins next week, she should shadow you on engagement meetings and start wire framing the content as she is learning," Gary noted. "With her solution architecture background, she'll be quick to adapt these cloud use cases and develop technical content for these roadshows that's accessible to non-specialists."

Gary began gathering his notes. "I'll share our plans with Charles next week. This 'Governance Roadshow' approach aligns perfectly with his vision for cloud enablement. By investing in educating the organization about our cloud capabilities, we're not just driving adoption, we're positioning Goliath to more effectively leverage modern IT practices and stay competitive."

As they wrapped up their discussion, Gary felt energized by the proactive direction they were taking. This wasn't just about building better governance tools anymore; it was about bringing the entire organization along on their cloud journey.

Day 469: Audit Progress

"I never thought I'd say this, but I'm actually looking forward to our audit review," Gary said, arranging documents on the conference room table. "Isabel's going to be surprised by how far we've come."

What had once been a source of anxiety was now approached with confidence, reflecting the significant progress they'd made since her initial findings.

Sarah, who had helped him prepare for this discussion with Isabel, was sitting next to Gary. "These numbers speak for themselves. Our account coverage increased significantly, folding in more cloud accounts into our process. With the growing compliance adherence percentage of our controls, and the expansion of new controls to increase our IAPP coverage, we are in good shape for a positive review with Audit."

"And it's not just the numbers," Gary added. "It's the systematic approach. We've established a framework that can demonstrate continuous improvement, not just point-in-time fixes."

The governance platform dashboard displayed a visible progression of their control implementation. Each phase of their rollout approach was visible, showing how controls moved from preview to check to enforcement across different environments. Just as importantly, it highlighted where controls were muted or exceptions had been granted, with a detailed audit history.

As Isabel arrived, Gary sensed a different dynamic from their previous meetings. There was still professional scrutiny, but the hostility that often characterized Audit interactions had given way to something closer to partnership.

"Before we begin," Isabel said as she settled into her seat, "I want to acknowledge the progress your team has made since our last assessment. The monthly updates you've been providing have shown steady improvement in addressing our initial findings and continuing to grow the coverage of IAPP controls."

"Thank you," Gary replied. "We've taken your recommendations seriously, used them as a foundation for building a sustainable governance approach, and made investments to mature our processes with additional capacity and tooling."

Isabel nodded. "Let's start with the control adherence, which was my primary concern from the last review."

For the next hour, Gary and Sarah walked through their governance framework, demonstrating how they identified which controls to implement, and how those were deployed and tracked. Isabel asked pointed questions and received thorough, evidence-based answers backed by metrics and documentation.

"I appreciate how you've started with a foundational approach to satisfy core security-related IAPP controls by mapping industry-accepted CIS Benchmarks in your cloud framework." Isabel noted, reviewing the dashboard.

"The mapping has gone a long way," Sarah acknowledged, "it's been valuable beyond just audit compliance. It helps teams understand how technical controls relate to organizational requirements."

"I've noticed you also introduced financial and operational controls?" Isabel asked.

"While our security controls remain critical, we've extended our framework to bring forward financial controls to help with a management initiative to lower IT expenses, and they map well to increase our IAPP coverage," Sarah explained, showing examples. "Each control mapped back to specific IAPP policies and includes the same implementation guidance and verification methodology."

Isabel nodded, "These are great additions to the framework and great for the breadth of coverage coming along." She added, "I would like to better understand the process of handling deviations and controlling exceptions."

Gary demonstrated their tiered exception process, explaining the distinction between temporary operational muting and formal policy exceptions. "Muting suppresses alerts while preserving the underlying security requirements. Actual exceptions modify the policy itself and therefore require more rigorous review and approval."

"That's a nuanced approach," Isabel observed. "It acknowledges operational realities while maintaining control integrity."

As the meeting progressed, Isabel's questions became less about compliance gaps and more about how certain approaches could be extended to other areas of the organization.

"The approach you have implemented with continuously improving your coverage, communicating out upcoming controls, assessing exceptions, and tracking compliance could be valuable beyond just cloud environments," she noted. "We have similar challenges with our traditional infrastructure process and the approach you have taken could be a welcome improvement to consider standardizing for on-premises controls."

By the meeting's conclusion, Isabel's assessment was decisive: "Based on what I've seen today, you've addressed all critical findings from our previous audit.

There are still some moderate recommendations to work through, but the foundation is solid. As noted early on, we still expect the team to continue increasing coverage of IAPP controls, but we are more confident you have the foundation to make that growth achievable."

She continued, "What's most impressive isn't just the technical implementation, but how you've integrated governance into the broader operational processes. This isn't just a security or financial process, it's a foundational capability for Shared Services supporting those practices."

After Isabel departed, Gary turned to Sarah with a smile. "This is a significant milestone. A year ago, we were struggling just to identify our cloud accounts. Now we have a governance framework that's being recognized as a model for the organization."

"It's not just about appeasing our Audit team," Sarah replied. "It's about building capabilities that actually make cloud adoption safer and more efficient for Goliath."

Gary nodded in agreement. "This is what effective cloud governance looks like. Not a reactive response to incidents or upcoming audits, it's a proactive, supportive and systematic framework that is embedded in how the organization operates."

Day 476: Charles's Evolution

Twenty-eight years at Goliath Federal Bank, and Charles was still learning new ways to approach technology leadership.

Charles reviewed the quarterly reports on his desk, traditional infrastructure metrics alongside Gary's cloud governance dashboard. The contrast was striking. The traditional PowerPoint presentation reflected Goliath's methodical, risk-averse approach, while the cloud metrics showed something different: adaptive decision-making, detailed real-time insights, and continuous improvement.

When the board had first approved Goliath's cloud strategy three years ago, Charles had approached it like any other infrastructure initiative. They just treated it like a co-location, another way to provision servers, not realizing the transformational impact cloud would have on how Shared Services could operate in a world where infrastructure is now code, everything is an API, and automation is key to staying ahead.

The cloud wasn't just another datacenter. It was a fundamental shift that required much more investment in transformation. The organization had a knowledge and skillset gap that needed a more nuanced understanding beyond simply lifting and shifting workloads.

Charles reflected on hiring Gary as Cloud Architect nearly 500 days ago. It had seemed like a straightforward architecture role to support cloud projects and find patterns. But Charles now realized how much he had underestimated the scope of what was needed.

His own management approach had been too passive in those early days. He'd given Gary space to find his way but hadn't invested fully in defining a strategy or ensuring success. From that disastrous town hall to the struggles with internal politics, Charles could have provided more support and mentorship.

Fortunately, Gary had been the right hire, overcoming adversity and finding the right internal partners. He'd built momentum through major moments like security incidents, audit findings, and cost-saving initiatives that added urgency along the way.

Charles had watched something different unfold as Gary built relationships across divisions, engaged with business stakeholders, and gradually shifted the conversation from control to enablement. The watershed moment had come during the security incident with Jason. What could have been a standard crisis management exercise became the catalyst for real organizational change. Instead of only fixing the immediate problem, Gary had used it to illuminate systemic issues and build cross-functional alignment.

Charles picked up the governance dashboard, studying the metrics. Account compliance rates, control implementation status, exception approvals, remediation timelines; all the technical indicators showed steady improvement. But the real story was in the engagement metrics: increasing voluntary adoption, broader participation in the Cloud Champions network, growing attendance at training sessions, new application and business teams reaching out asking how to solve business problems together.

These metrics reflected a significant cultural shift in how Goliath approached cloud technology. Cloud wasn't just another IT service offering, it was a business capability.

The shift to cloud required a complete transformation at Goliath. It had forced Shared Services to reorganize and unify around a cloud governance process, and to think differently about their user experience: from request fulfillment to self-service with automated control.

The shift to governance was also a shift in the culture. Charles had begun using a metaphor that seemed to resonate with both technical and business leaders: "Cloud governance isn't about building walls; it's about paved roads with guardrails. We're not trying to restrict movement; we're creating safe, efficient pathways that help everyone reach their destinations faster."

The upcoming CloudNexis presentation was an opportunity to share this perspective more broadly. Having Goliath present its governance approach at a major industry conference positioned the bank not just as an adopter of cloud technology, but as a thought leader in how to enable cloud services faster and safely.

Charles gathered his materials, satisfied with their evolution. They now had a customer-centric service with continuous improvement built in with visibility to understand what was happening, automated operational tasks to reduce noise, and levers to pull when needs arose, like cost-cutting measures.

It was a complete transformational shift he was glad to support. While he might reflect on how they could have done it differently, he couldn't imagine operating the old way now. The journey had made him a much better IT leader.

Day 483: Presentation Preparation

"How to Herd Clouds and Influence People," Gary muttered, staring at the title slide on his laptop. The conference was just two weeks away, and as he reviewed his presentation, something felt off about his approach.

His phone buzzed with a calendar reminder: a 1:1 with Charles in ten minutes.

Charles asked about Elena's progress and their latest control rollouts before turning to the presentation. "How's the CloudNexis presentation coming along?"

Gary hesitated. "I submitted the initial slides last week after I got approval from Communications, but I'm having second thoughts about the whole approach."

"What doesn't feel right about it?" Charles asked.

"I structured it as Goliath's chronological journey; from my joining through the governance platform implementation," Gary explained. "But it feels too focused on what we did, not what others can learn."

Gary continued, "I started collecting stakeholder perspectives to make sure everyone's contributions were represented. Barry wants more emphasis on

Commercial Banking's early cloud adoption. Dennis wants to highlight the governance evolution based on his feedback. It quickly became about managing internal politics rather than offering genuine insights."

Charles smiled knowingly. "Presentations like this often become political negotiations. Everyone wants their role highlighted. But that's rarely what makes them valuable to external audiences."

"Exactly," Gary agreed. "Other organizations don't care about our org chart or when we formed committees. They want to understand how we approached the problem and why certain strategies worked or failed."

"When I present to the Board, I never just give them a timeline," Charles noted. "I focus on insights, implications, and strategic direction. The milestones are just illustrative examples."

"If I were in your shoes, I'd organize around core principles that any organization could apply, using our experiences as examples that support those principles."

Gary considered this. "Like the framework we've been developing internally: 'Know Your Cloud,' 'Raise the Bar,' 'Make Change Happen.' That's more universally applicable than details of our journey."

"Precisely," Charles replied. "Position Goliath as a thought leader by educating others on how to bring Cloud Governance principles to their organizations. That doesn't come from telling people what committees you formed. It comes from articulating a framework that helps them think differently about their own challenges."

"Think about the business books that have lasting impact," Charles noted while looking at his bookshelf. "They extract principles that can be widely applied, not just tell one company's story."

"If only folks could read about transforming your strongest opponent into your most vocal supporter," Gary said with a laugh, thinking of Barry's evolution.

"There's a Barry in every organization," Charles replied with a smile. "They're the toughest to win over, but once they're on your side, they become your strongest advocates."

"Thanks, Charles, I appreciate your advice. I am going to take another crack at it, focusing on our lessons learned."

Back in his office, Gary started fresh, restructuring his presentation around the governance principles they'd developed. Instead of organizing by Goliath's timeline, he focused on the universal challenges of cloud governance and how their framework addressed them.

CloudNexis would be an opportunity not just to share Goliath's accomplishments, but to advance the practice of cloud governance itself.

Day 497: Clouds on the Horizon

The sprawling expo floor of CloudNexis buzzed with activity as Gary and the Goliath contingent dispersed to explore. Charles had already arrived a day early for the executive track, but had secured conference passes for a surprisingly large group from Shared Services: Mark, Sarah, Fran, Oscar, Adam, Nancy, and Isaac. And application teams got the budget to attend as well. Gary knew Peter, Dennis, Mike, Barry, Jennifer, Robert, and Eric were all traveling. It was a shame Elena couldn't make it, but Goliath was well represented.

"Let's meet at the Financial Services Luncheon at noon," Gary suggested. "That gives everyone a few hours to explore the floor, get lunch at the event, and make it to the first afternoon session."

As the team scattered, Gary found himself drawn to the huge vendor booths with impressive displays but vague messaging, and to the smaller booths trying to stand out with interesting hooks. But noticeably, cloud security, FinOps, and AI were sprinkled into everyone's messaging. What struck him was how little emphasis there was on governance as a unifying concept. Vendors focused on point solutions, security tools, cost management platforms, and operational monitoring, but few discussed how these elements came together.

Unlike the rest of the team, who could freely explore, Gary's morning was already scheduled solid with pre-arranged meetings. His role as "Director, Cloud Governance" had made him a target for vendors wanting to discuss their governance solutions and platform providers eager to showcase their latest features. The Goliath name on his badge further ensured his popularity.

Walking between meetings, Gary passed Mike with an overflowing bag of swag from random vendors. He was clearly going for the “most swag collected” from the team.

"Having fun?" Gary asked, gesturing at the overflowing bag.

"Going for the record," Mike grinned. "The kids expect a haul when Dad goes to a conference."

As Gary navigated the expo floor, he spotted Sarah engaged in conversation at a security vendor's booth and Fran deeply involved in a discussion at the FinOps Foundation area. Everyone seemed to have found their niche.

By noon, Gary had collected several business cards and speaking invitations from organizations eager to learn more about Goliath's approach.

Walking toward the Financial Services Luncheon, Gary spotted Barry surrounded by IT executives from other banks, gesturing animatedly as he spoke.

"We pioneered cloud adoption at Goliath," Barry was saying. "My Commercial Banking organization saw the potential early, while central IT was still focused on data centers. Once we demonstrated the value, others followed our lead."

Gary smiled to himself. Barry wasn't entirely wrong, though he was certainly polishing the narrative to put his division at the center.

"And how did you handle being slowed down by regulatory challenges?" someone asked.

"We developed a pragmatic approach," Barry replied. "Standards that enable innovation rather than restrict it. We've found that implementing a governance framework works best when it helps teams move faster safely, not just saying 'no' faster."

Gary almost laughed out loud. Barry was now advocating the very philosophy he'd once resisted. Yet there was something satisfying about hearing him internalize and champion these ideas as his own.

After some time mingling at the Financial Services Luncheon, the Goliath team grabbed a table together to share their morning observations.

"What did you think of the expo floor?" Gary asked the group.

"It felt like speed dating at the FinOps Foundation booth," Fran replied. "I spent most of my morning there. The common theme? Everyone is struggling with accurate reporting and showbacks. Many organizations can track costs but can't tie them back to business functions or get teams to take ownership."

"I was comparing notes with another security lead from Maplewood Bank," Sarah added. "They're finding that teams are willing to secure new workloads but struggle to fix legacy configurations. Sounds familiar."

"What about you, Barry?" Gary asked. "I saw you had quite the audience earlier."

Barry straightened slightly. "There's significant interest in how we are managing our cloud environments. I've arranged a private discussion with Oakview National's leadership team after your presentation tomorrow."

Gary expressed his appreciation. "That's great. Their perspective would be valuable."

As lunch concluded, Peter pulled Gary aside. "You know what I'm finding most interesting? Every division is engaged with completely different aspects of cloud governance, yet they're all seeing value through their own lens."

Gary considered this insight. "A year ago, most of them viewed what we now call governance as something imposed on them. Now they each see it as a capability that serves their specific needs."

Gary checked his watch. "I've got to head to the financial services executive roundtable. It was on my calendar for weeks, and I was hoping to catch at least one regular session today."

"No worries," Dennis replied. "The rest of us are hitting the multi-cloud strategy session. We'll compare notes at dinner."

The financial services roundtable turned out to be a candid discussion where executives shared their cloud challenges.

"Regulatory expectations keep increasing while the technology evolves even faster," one CIO explained. "How do you keep pace without freezing innovation?"

"Shadow IT is our biggest headache," another offered. "Teams using department credit cards to spin up cloud resources with no oversight."

"Talent is our constraint," a third added. "Finding people who understand both cloud technology and financial services requirements seems nearly impossible."

As Gary listened, he realized these organizations were describing exactly the issues their framework addressed. They all had tools for visibility, knowing their cloud, but were struggling with the next steps to raise the bar through systematic controls and make change happen through organizational adoption.

When asked about Goliath's approach, Gary briefly outlined their three-pillar framework. The executives exchanged glances, seemingly surprised by the simplicity yet comprehensiveness of the model.

"That's what's missing from our approach," one admitted. "We have all these separate initiatives around security, cost, and operations, but no unifying framework to bring them together."

As Gary walked back to his hotel room, he reflected on the day's conversations. Every vendor booth promised better visibility, more comprehensive scanning, smarter analytics, deeper insights into cloud environments. The financial services roundtable had been filled with executives talking about problems, but not how they were overcoming them.

But what struck him was how little focus there was on the harder problem: actually getting things fixed. Organizations were investing heavily in CNAPP tools that generated thousands of security findings, FinOps platforms that identified millions in potential savings, and operational monitoring that highlighted performance issues. Everyone was buying lists of problems. The real challenge wasn't finding issues, it was systematically addressing them across hundreds of accounts and dozens of teams. It was building the organizational capability to translate visibility into action, alerts into remediation, and recommendations into real change.

Most conversations focused on detection capabilities: "Our tool finds 30% more vulnerabilities than theirs." But Gary hadn't heard anyone discussing remediation rates, organizational change management, or how to scale governance across diverse teams with competing priorities.

That was the gap Goliath had learned to bridge; not just knowing what was wrong, but building the framework to systematically make it right.

Tomorrow, he would take the stage to share that framework with the industry. Today had validated they were on the right track and Gary felt more confident that his session would resonate with others.

Day 498: The Presentation

Gary stood at the side of the stage, reviewing his speaker notes one last time. In moments, he would step into the spotlight to deliver a presentation he hoped would strike the right balance between governance principles and practical examples.

The presentation wasn't the end of their journey but a milestone along a continuing path. For Gary, this felt like a transition from a career focused on execution to leadership. From his days at his prior company executing in a silo without organizational complexities, to building a capability for a massive decentralized organization, to now sharing wisdom with others beyond the walls of his organization.

Taking a moment to look out at the audience, Gary could see the room was packed. People stood along the back wall and sat in the aisles. In the front rows, he spotted his Goliath colleagues. Mark and Sarah sat attentively, notebooks open. Fran was beside them, along with Oscar, Adam, Nancy and Isaac, all chatting amongst themselves with smiles on their faces.

Dennis gave a subtle thumbs-up from the middle of the room. Peter sat nearby, whose journey had paralleled the governance evolution at Goliath. And there, surprisingly close to the front, sat Barry. After all their battles, all the resistance, all the territorial disputes, Barry had positioned himself prominently for this presentation.

Charles took a seat in the back. Partly because he was coming late from an executive meeting, but Gary sensed it was also his way of giving him space for this moment.

"Our next speaker has led the cloud transformation at Goliath Federal Bank," the moderator was saying. "Please welcome Gary Anderson, Cloud Governance Director, to share insights on 'How to Herd Clouds and Influence People.'"

Taking a deep breath, Gary walked onto the stage.

Forty-five minutes later, Gary concluded with the words he'd rehearsed but now delivered with genuine conviction: "This is Governance for the Cloud Age."

The applause was immediate and sustained. Several audience members rose to their feet. The response exceeded his expectations.

During the Q&A that followed, the questions revealed how deeply the framework resonated. A CTO from a healthcare company asked about adapting the approach for their specific regulatory requirements. A startup founder wanted to know how to implement governance without stifling innovation. A fellow bank executive inquired about the ROI of governance platforms.

But the question that stuck with Gary came from a developer in the back: "How do you get teams to actually want governance instead of just tolerating it?"

Gary smiled, thinking of Dennis volunteering for pilots, Peter's enthusiastic adoption, even Barry's evolution. "You stop trying to control them and start trying to enable them. When governance makes their lives easier instead of harder, they'll ask for more of it."

As the session ended and people began filing out, Gary was approached by a steady stream of conference attendees. Business cards were exchanged, follow-up meetings arranged, and invitations extended. The presentation had clearly struck a chord in the industry.

* * *

That evening, the Goliath team gathered at a restaurant near the conference center. Charles had managed to secure a private dining room, and the entire group was present, including Barry.

"To Gary," Charles raised his glass as appetizers arrived, "for not only a great presentation today, but for building something remarkable at Goliath."

"To the team," Gary quickly replied feeling embarrassed, raising his own glass. "This journey didn't emerge from one person's vision. It came from all of us figuring it out together."

As the last glasses clinked, Dennis said grinning, "I have to admit, when you first showed up talking about cost optimization scripts, I thought you were just another central IT person who didn't understand how we actually work."

"You weren't wrong," Gary laughed. "I had no idea what I was walking into."

"The town hall," Sarah added with a wince. "That was painful to watch."

"But necessary," Peter observed. "Sometimes you have to fail spectacularly before you can succeed meaningfully."

"Speaking of spectacular failures," Oscar said with a grin, "remember when we were tracking everything in spreadsheets?"

The table erupted in groans and laughter.

"Or when Gary and I were manually reconciling AWS bills with expense reports," Fran chimed in.

"'Developmet,'" Mark said, making air quotes. "One typo that nearly broke our entire policy framework."

"But we fixed it in hours instead of days," Gary pointed out. "That's progress. And honestly, going through all those manual processes taught us what we really needed to automate."

"Like the governance platform," Isaac noted. "Suddenly we could implement controls in days instead of months."

"And teams could fix issues in minutes instead of submitting tickets," Dennis added.

Gary turned to Barry, who had been quietly listening with an amused expression. "Barry, you've seen this evolution from a different perspective. What changed your mind about working with us?"

Barry paused, swirling his wine. "You know, I spent twenty years successfully setting our own paths in Commercial Banking. I was good at it." He looked around the table. "What changed my mind wasn't the technology or even the business case. It was realizing that you were actually trying to make our lives easier, not harder."

"That wasn't always the case," Gary admitted. "I definitely started with a controlling mindset."

"We all did," Sarah added. "I was sending security alerts like they were all critically urgent. No wonder teams ignored them."

"The breakthrough came when we stopped asking 'How do we make teams comply?' and started asking 'How do we help teams succeed?'" Mark reflected.

Fran nodded enthusiastically. "The difference between showing teams their problems and giving them tools to solve their problems."

"You know what I love most about this?" Charles said. "How it became this multiplier effect. The more capabilities we built, the more teams wanted to engage. Success really did breed success."

"The Cloud Champions network was brilliant," Jennifer from Retail Banking added. "Having advocates in each division made all the difference."

"It wasn't planned," Gary admitted. "Those advocates emerged naturally. We just recognized them and gave them a platform." He glanced at Barry with a slight smile. "Like Barry. From our biggest skeptic to our most vocal supporter."

Barry raised his glass. "Someone had to keep you honest. But once you proved you could deliver value without creating bureaucracy, I was all in."

"You know what I'm most proud of?" Gary said. "The change in how we work together. A year ago, this dinner wouldn't have happened. Different divisions didn't collaborate. Central IT was seen as a roadblock."

"Now we have teams asking for more governance," Sarah marveled. "That's a sentence I never thought I'd say."

"Because governance isn't the enemy of innovation," Fran observed. "Done right, it's what enables innovation at scale."

"Remember when we were actually excited for our audit review with Isabel?" Sarah laughed. "I never thought I'd look forward to a meeting with Internal Audit, but we'd built something we were genuinely proud to show her."

"Your presentation today really captured our journey," Charles said thoughtfully. "You articulated that the biggest challenge in governing clouds isn't the technology. Visibility is critical to knowing what's wrong, but how you effectively improve and raise the bar and make change happen is really the challenge you captured well."

"What's next?" Charles asked. "After tonight, there will be a lot of interest in what we've built."

Gary considered the question. "We keep improving. The cloud never stops evolving, and neither can we. New services, new threats, new business needs. Our job is to embrace that pace of innovation, not try to slow it down."

"And we help others," Mark added. "Share what we've learned, learn from what others are doing."

"Maybe write a book," Dennis suggested with a grin. "How to Herd Clouds and Influence People."

"I think one presentation is enough for now," Gary laughed.

As the evening wound down and team members began to head back to their hotels, Gary found himself reflecting on the journey. From that first day when he couldn't even log in, to tonight's celebration of an approach that could help organizations worldwide, it had been quite a ride.

"Penny for your thoughts?" Sarah asked as they waited for the elevator.

"Just thinking about how far we've come," Gary replied. "Not just with the technology, but with each other. This team, this approach, it's something special we've built."

"It is," she agreed. "And it's sustainable. That might be the most important part. We didn't just solve immediate problems; we built capabilities that will keep improving things as we grow and evolve."

As the elevator arrived and the team began to disperse, Gary felt a deep sense of satisfaction. Tomorrow they would return to Goliath and continue the work: new controls to implement, new teams to support, new challenges to address. But tonight, they had proven that cloud governance could be something more than a necessary evil. It could be a competitive advantage, a source of innovation, and a way to bring people together around common goals.

The presentation had gone well, but the real success was the team that had built this capability together. That was the story worth telling.

Appendix: Gary's Presentation

The following is the presentation Gary delivered at CloudNexis.

Slide 1: "How to Herd Clouds and Influence People"

"Thank you for that introduction," Gary began, then paused, taking in the packed room. "I have to admit, I'm a bit surprised to see so many people interested in a talk about herding clouds. I thought everyone might have mistaken this for the AI session down the hall."

A ripple of knowing laughter spread through the audience, easing his nerves.

"But the fact that you're all here tells me something important. It tells me that organizations across industries are recognizing that managing cloud environments effectively is one of the most significant challenges we face today."

Gary noticed several people nodding in agreement. This was clearly a shared pain point.

"Based on the conversations I've had at this conference, it's clear that many of us are wrestling with the same challenge: how to manage cloud environments effectively without stifling the innovation and agility that drew us to cloud in the first place."

Slide 2: Cloud Complexity at Scale

"As organizations adopt cloud at scale, they face complexities that outpace their ability to address them," Gary continued. "The challenges are multifaceted and growing."

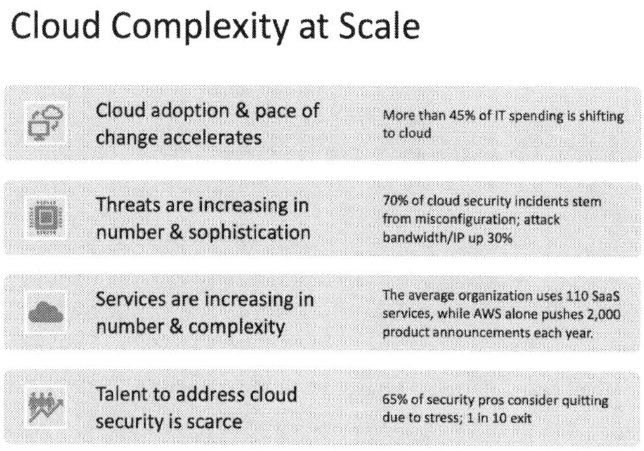

"These trends create a perfect storm for enterprises," Gary explained. "The complexity and pace of change are simply outpacing most organizations' ability to address it effectively. Just keeping up with the latest services and features is a full-time job, let alone securing and optimizing them."

Gary pointed to a chart showing the exponential growth in cloud services over the past five years.

"And it's not slowing down. The major cloud providers are innovating faster than ever, which creates both opportunities and challenges for organizations trying to manage these environments."

Slide 3: Cloud Changed How Organizations Work

"While these technical complexities are significant, there's an even bigger challenge that many organizations overlook," Gary continued, advancing to the next slide. "The organizational shift that cloud demands is often more difficult to navigate than the technical challenges themselves."

"Cloud fundamentally changed how organizations operate," he explained, pointing to the transformation diagram on the slide.

"This isn't just a technology change; it's a fundamental shift in organizational power and responsibility," Gary emphasized. "Infrastructure became code, which means it became part of the application development process. App teams now have direct control over the infrastructure their applications depend on."

"But here's what many organizations miss," Gary said, highlighting the challenge. "This shift happened whether they planned for it or not. App teams gained unprecedented agility and control, but most Shared Services organizations haven't adapted their operating model to match this new reality."

He clicked to show the common organizational response.

Cloud Changed How Organizations Work

Traditional IT	Cloud-Native IT
Request/fulfillment with centralized control	Infrastructure as code managed by application teams
Shared Services owns & manages infrastructure	Self-service platforms with guardrails
Physical barriers and approval gates	App teams have direct infrastructure control
Clear separation: app vs. infrastructure teams	Shared Services provides foundations & oversight

"The typical reaction is to try to impose traditional controls on cloud environments. Organizations port their existing on-premises processes: approval workflows, manual reviews, centralized provisioning. And mistakenly apply them to cloud resources."

"This approach completely corrupts the innovation and agility benefits of cloud while trying to maintain control mechanisms designed for a different operational model."

Gary pointed to the tension illustrated on the slide.

"The result is organizational conflict. App teams want to move fast and leverage cloud capabilities. Shared Services teams want to maintain control and ensure governance. Without adapting the operating model, these goals seem incompatible."

"Most organizations focus on solving the technical complexity," Gary continued, "but they underestimate or ignore the organizational transformation required. They invest in tools and processes without addressing the fundamental shift in how teams work together."

"The solution requires us in Shared Services to completely rethink our role," Gary concluded. "Instead of controlling what app teams do, we enable them to do it safely. Instead of gates that slow teams down, we provide guardrails that keep them within safe boundaries while preserving their autonomy."

"This organizational shift, from controlling to enabling, is often the hardest part of cloud transformation. But it's also the most critical for long-term success."

Slide 4: Current Approaches and Their Limitations

"So how are organizations approaching this challenge today?" Gary asked, gesturing to the slide that illustrated current approaches.

"Most organizations are implementing fragmented solutions," he continued. "Shared Services teams work in silos with different tools and priorities. There might be a small Cloud Team focused primarily on account brokering and billing management, but no central driving force optimizing the health and posture of cloud services."

The slide showed the typical organizational response.

Current Approaches & Their Limitations

Fragmented Shared Services

- Security: CNAPP tools
- Finance: FinOps
- Ops: monitoring
- Network: connectivity
- Identity: access

Limited Cloud Team investment

- Focus on account provisioning & cost tracking; lacks authority for org-wide improvement

Outcomes

- Alert overload; conflicting priorities; little guidance on what to fix first or how

"Application teams are left to navigate between these different requirements," Gary explained. "They get security alerts from one team, cost recommendations from another, operational guidance from a third. Each team speaks their own language and has their own priorities."

He pointed to a diagram showing the overwhelming flow of disconnected requirements hitting application teams.

"The result is that we've invested heavily in visibility tools that are excellent at finding issues, but we've created a flood of alerts that overwhelm teams," Gary continued. "Security tools generate hundreds of findings. FinOps platforms identify dozens of optimization opportunities. Monitoring solutions highlight performance issues."

"Teams can't distinguish critical issues from minor ones. They don't have defined guidance on what to fix first, how to fix it, or why it matters to their specific context."

Gary clicked on a visual showing typical notification overload from various tools.

"Look familiar?" Gary asked, noticing nods throughout the audience. "These tools are all valuable individually, but together they create an overwhelming flood of information that paralyzes rather than enables action."

"The visibility is there, but teams are drowning in notifications without the organizational framework to translate that visibility into systematic improvement."

Slide 5: Defining Governance

"The issue is that we have been rushing to keep pace with the cloud, bolting on tools and fragmented processes. We need to take a step back and frame what's missing; how do we actually manage this at scale? This is where governance becomes essential," Gary explained. "But what exactly do we mean by governance?"

The slide displayed the core elements of governance:

Defining Governance

- **Rules/Laws**: Framework & policies
- **Oversight**: Audit & judgment
- **Management**: Infrastructure & process
- **Protection**: From threats
- **Transparency**: Trust, visibility & communication

"At its core, governance provides the structure that enables freedom," Gary emphasized. "It's like traffic laws; they don't exist to prevent you from driving, but to ensure everyone can drive themselves safely and effectively. In a way, that is what we want from our application teams using the cloud. I want them in full control to get to their destination without hindering their ability to just get going."

"Effective governance isn't about restrictions, it's about creating guardrails that enable safe innovation. It's about establishing defined boundaries within which teams can operate with maximum autonomy and minimum friction."

Slide 6: Traditional vs. Cloud Governance

"But we can't just lift and shift our traditional IT governance for the cloud," Gary continued, showing a comparison table on the slide.

"If we simply carry over existing processes to the cloud, we corrupt the very innovations and agility that drew us there in the first place," Gary explained. "The cloud demands a new governance model that preserves agility while managing risk appropriately."

He advanced to a slide showing examples of how traditional governance models fail in cloud environments.

Traditional vs. Cloud Governance

Element	Traditional Governance	Cloud Governance
Rules/Laws	Frameworks, policies, checklists	Policies as code
Oversight	Lengthy arch reviews, assessments, audits	Real-time evaluation via controls
Management	Request/fulfillment processes	Self-service with automation
Protection	Physical barriers/controlled usage	Preventative & automated corrective controls
Transparency	Siloed visibility & comms	Real-time, accessible change control & guidance

"Traditional governance assumes static infrastructure and infrequent changes. But in the cloud, resources are ephemeral, configurations change constantly, and teams deploy multiple times per day. The governance model needs to adapt to this dynamic reality."

"This shift in governance approach is what enables the three-pillar framework I want to share with you. Each pillar addresses a fundamental challenge that traditional governance models struggle with in cloud environments."

Slide 7: The Governance Framework

"This brings us to the framework we've developed at Goliath," Gary said, revealing the three pillars on the slide. "It addresses the complexities we've discussed while balancing control with enablement."

He walked through each pillar:

- **Know Your Cloud**: Establishing comprehensive visibility and insights.
- **Raise the Bar**: Implementing controls that enable improvements.
- **Make Change Happen**: Driving adoption through organizational transformation.

"This framework doesn't just address technical challenges, it recognizes that effective cloud governance is as much about people and process as it is about technology."

Gary noticed several attendees taking photos of this slide with their phones. The simplicity of the framework seemed to resonate immediately.

Slide 8: Know Your Cloud, Why It's Difficult

"Let's dive deeper into the first pillar: 'Know Your Cloud,'" Gary began. "I've had numerous conversations here at CloudNexis where visibility challenges come up repeatedly. So before I talk about our approach, let's look at why achieving meaningful visibility is so challenging."

The slide showed three major obstacles:

- Siloed Access: Different teams see different pieces of the puzzle
- Missing Context: Alerts lack business and operational context
- Prioritization Chaos: No framework to balance competing priorities

"While most organizations have visibility tools in place, they still struggle to translate that visibility into actionable insights," Gary explained, gesturing to a diagram showing fragmented tool access.

"First, there's the access problem. Who actually has access to these tools? Often, access is siloed within specialized teams: security teams see security data, finance teams see cost data, operations teams see performance data. No one sees the complete picture.

"Even when we solve the access problem, there's the context problem," he continued, showing an example of a security alert without contextual information. "Alerts lack the business and environmental context needed to evaluate their importance. Is this S3 bucket misconfiguration affecting a production system or a test environment? Is that cost spike related to a critical business initiative or an abandoned experiment? Without this context, teams can't make informed decisions about what to prioritize."

"And perhaps most challenging is the prioritization problem," Gary said, clicking to a complex prioritization matrix. "With thousands of findings across different dimensions, how do you decide what matters most? How do you balance a medium security risk against a high cost optimization opportunity? Most organizations lack a framework for making these decisions consistently."

He displayed a final screenshot showing alert overload across multiple dashboards.

"The result is what I call 'visibility without action.' We know there are problems, but we lack the coordination, context, and prioritization framework to address them effectively. Teams end up drowning in alerts while actual risks remain unaddressed."

"So how do we address these challenges?"

Slide 9: Know Your Cloud, Our Approach

"Our approach to 'Know Your Cloud' isn't just about having monitoring tools," Gary explained.

"It's about transforming fragmented data into actionable intelligence that enables better decisions. We're not lacking data, we're lacking the integration and context that makes that data useful."

The slide displayed three critical aspects of cloud visibility:

- Security Posture (CNAPP): Vulnerabilities, misconfigurations, compliance gaps
- Financial Management (FinOps): Cost trends, optimization opportunities, allocation accuracy
- Operational Health (Observability): Performance, reliability, resource efficiency

"The key breakthrough for us wasn't implementing better monitoring tools, it was translating the recommendations into controls that connect cloud resources to business capabilities and organizational responsibility."

Gary highlighted a visual showing how Goliath had evolved from struggling to identify cloud accounts to establishing comprehensive visibility.

"Effective visibility answers critical questions: What resources do we have? Who owns them? What business functions do they support? What risks do they pose? What do they cost? How are they performing? What's changing in my environment?"

"Without this foundation of visibility of knowing your cloud, any attempt to implement controls or drive change will be built on quicksand. You can't improve what you don't understand, and you can't manage what you can't see."

Slide 10: Raise the Bar, Why It's Difficult

"Once we have visibility, the next challenge is actually fixing the issues we identify," Gary continued. "I'm betting many of you have experienced this frustration where you know what's wrong, but it's like moving mountains to get anything fixed between teams."

The slide illustrated this difficulty:

- The Remediation Maze: Complex workflows with multiple handoffs
- The Prioritization Problem: Feature development vs. infrastructure improvements
- The Expertise Gap: Specialized knowledge required across multiple domains
- The Overhead Burden: Tracking, meetings, escalations consuming more time than fixes

"The traditional approach creates an operations flow nightmare," Gary explained, tracing a finger along the complex workflow diagram. "We either bundle findings into major remediation projects that never get funded, or we push individual alerts to application teams who are already overwhelmed with feature development and other priorities."

"Let me walk you through what this typically looks like," Gary said, clicking to an animated flowchart showing the life of a security finding." A security tool identifies a misconfiguration. It generates an alert. The security team reviews it, creates a ticket, and assigns it to an application team. The application team doesn't understand the issue, so they request clarification. Security responds. The application team prioritizes it below feature work. Security escalates after a deadline passes. Management gets involved. Meetings are scheduled. Finally, weeks later, a fix is implemented. But then a similar issue appears elsewhere, and the cycle begins again."

"Think about the overhead involved," Gary continued, highlighting the inefficiencies in the process. "For each finding, there are emails, tickets, meetings, follow-ups, escalations, status reports. We're spending more time tracking and discussing issues than actually fixing them."

"This is why we don't see major improvements even with all our visibility tools. It's not just a technology problem, it's a data, process, and people problem."

"So how do we break this cycle?"

Slide 11: Raise the Bar, Our Approach

"Our approach to 'Raise the Bar' focuses on implementing controls that systematically improve your posture," Gary explained. "But the approach matters as much as the controls themselves."

The slide displayed the key components of effective control implementation:

- **Control Lifecycle**: Draft → Preview → Check → Enforce
- **Control Types**: Cost optimization, security configuration, operational reliability

- **Control Enablement**: Self-service remediation, automation, exception handling

"We learned through trial and error that controls implemented without context or without consideration for operational impact are controls that will be resisted or circumvented," Gary continued. "Our rollout approach addresses this by moving controls through distinct phases."

He walked through the phases displayed on the slide:

- **Draft**: Internally testing and validating each control before sharing.
- **Preview**: Showing teams what will be evaluated before enforcement begins.
- **Check**: Monitoring compliance and providing self-service remediation tools.
- **Enforce**: Automating compliance where appropriate.

"This approach builds understanding and buy-in while providing teams time to adapt," Gary emphasized. "It transforms controls from barriers to enablers."

"The key insight here is that effective cloud governance isn't about preventing teams from doing things, it's about making it easier for them to do things the right way. The most effective controls are ones that reduce friction rather than creating it."

"We need to shift from alerts to managed controls. Controls that represent repeatable standards that are documented, communicated, deployed, and monitored. Standards that speak the internal language of the company and align with how we define our internal controls."

"A key insight from our experience was that most organizations have excellent visibility tools, but lack the workflow capabilities to systematically act on their findings. This gap between knowing what's wrong and fixing it systematically is where governance platforms become essential, as they bridge visibility to action through automated workflows."

Slide 12: Make Change Happen, Why It's Difficult

"Of course, implementing these approaches requires organizational change, which is perhaps the most challenging aspect of cloud governance," Gary explained, displaying a slide showing organizational silos and competing priorities.

The slide illustrated common organizational barriers:

- **Siloed Teams**: Different groups with different goals and incentives
- **Resistance to Change**: Natural human tendency to maintain status quo
- **Expertise Gaps**: Knowledge concentrated in pockets rather than distributed
- **Competing Priorities**: Business demands vs. infrastructure improvements

"How do you drive change in an organization where everyone is siloed?" Gary asked, gesturing to an organizational chart showing disconnected teams. "Not just in shared functions like security and finance, but across disparate application teams with their own goals and objectives?"

He clicked to show a graph of typical organizational resistance patterns.

"You could take a top-down approach, where management mandates the change," Gary explained, pointing to one path on the graph. "This creates urgency but often generates resistance and resentment. Teams comply minimally rather than embracing the change."

"Or you could rely on bottom-up adoption," he continued, indicating another path. "This builds buy-in but can take too long to achieve meaningful results, especially in large organizations where coordination is challenging."

"The reality is that we're human, and humans need time to adopt change, to be educated, to see it socialized and become the norm," he said. "Think back to when cloud was first introduced and how many felt it was a threat to their jobs. Or how DevOps and Agile methodologies were initially resisted as radical changes unsafe for delivering software."

"We're seeing similar reactions to AI today," Gary noted, drawing a parallel to current trends. "The initial reaction is often to dismiss and resist. We can't force change; we need to elevate our teams to adopt change and go along for the journey."

"So, how do we address these human challenges?"

Slide 13: Make Change Happen, Our Approach

"Our approach to 'Make Change Happen' addresses the human dimension of cloud governance," Gary explained. "This is where our framework moves from technical implementation to organizational transformation."

The slide showed key elements of effective organizational change:

- **Cloud Champions Network**: Distributed advocates across teams
- **Procedural Justice**: Fair, transparent processes for implementation
- **Education & Enablement**: Training, documentation, roadshows
- **Continuous Improvement**: Feedback loops, evolving patterns

"We've evolved from top-down mandates to collaborative enablement," Gary continued. "Our Cloud Champions network has been particularly effective, creating a community of practice that spans organizational boundaries. As more adopt the cloud, share their use cases, and provide feedback, it continues to build trust and credibility for the practice."

He highlighted how teams across the organization had become advocates for governance practices rather than resisters.

"The shift from enforcing compliance to enabling success is transformative," he emphasized. "When teams see governance as something that helps them rather than hinders them, adoption becomes natural rather than forced."

Gary described their Governance Roadshow approach and how they'd developed a library of resources to support teams at different stages of maturity.

"The most important lesson we've learned is that change happens at the speed of trust. By focusing on procedural justice, we are making implementation processes fair, transparent, and collaborative. We've transformed governance from something teams resist to something they request."

He pointed to a slide showing the evolution of their approach.

"We started by asking, 'How do we control the cloud?' Now we ask, 'How do we enable innovation while managing risk?' That shift in mindset changes everything."

Slide 14: Cloud Readiness Assessment

"As you think about applying these pillars in your own organizations, I'd like to share a simple readiness assessment. It's a set of questions that help diagnose where you are in your journey."

The slide displayed a series of questions organized by the framework pillars:

Know Your Cloud:

- Can you identify all cloud resources across your organization?
- Do you understand how cloud resources map to business capabilities?

- Do you have visibility into security, cost, and operational health in a unified way?
- Do you know who owns each resource and who to communicate with when issues arise?
- Do you have a framework for evaluating and prioritizing findings across security, cost, and operational recommendations?

Raise the Bar:

- Do you have a process for systematically implementing controls?
- Are your controls enabling teams rather than restricting them?
- Do you have mechanisms for handling exceptions and special cases?
- Do your controls map to internal standards for measuring progress?
- Do you have a process to identify which opportunities should be prioritized for the next control to deploy?

Make Change Happen:

- Do teams view cloud management as helpful rather than burdensome?
- Are teams actively participating in improving cloud practices?
- Do you have a collaborative forum for growing advocates and sharing use cases across business units?
- Do you actively communicate best practices and educate teams on cloud capabilities and governance benefits?
- Does your organization have a dedicated team or function to champion and mature cloud governance practices?

"These questions aren't just a checklist; they're a starting point for conversations about how governance can better serve your organization's needs. They help identify gaps and opportunities for improvement."

Gary could see people throughout the room contemplating these questions, mentally assessing their own organizations' readiness.

"I encourage you to use this assessment not as a scoring mechanism, but as a conversation starter with your teams. Understanding where you are is the first step to determining where you need to go."

"These assessment questions help you understand where you are, but the natural next question is, What do I do about it?" Gary said, clicking to the next slide.

Slide 15: Getting Started with Cloud Governance

"Let me share some practical guidance for getting started, based on where you are in your journey."

If you're just beginning:

- Start with visibility: use lightweight tooling to analyze your inventory, security posture, and usage. You can't govern what you don't know.
- Identify a cloud owner, ideally full-time, who can lead the transformation to bring together the framework.
- Partner with Security, Finance, Operations, Network, Identity, and other key teams to prioritize recommendations from your visibility tools and coordinate a shared implementation effort..
- Identify how recommendations align to your internal controls to build a common framework for measuring health.

If you have visibility but struggle with remediation:

- Always start small and iterate. Pick one high-impact, low-resistance recommendation to bring forward as a control to prove value.

- Implement a phased rollout with transparent communication. This brings teams along for the journey, sets the right expectations, and enables more successful adoption.
- Focus on self-service remediation to reduce organizational friction. The end goal is to avoid any manual touches as you build capability and trust for auto-remediation.
- Establish structured exception processes that balance flexibility with accountability.
- Have a model to track the success of control rollouts, exceptions, and how you're improving toward your defined standards.

"Our journey at Goliath actually began with just trying to figure out what cloud accounts our organization had," Gary explained. "Once we had that basic inventory, we could start making sense of the security and cost recommendations our tools were generating."

"As you start implementing controls systematically, you'll likely discover what we did; that the real challenge isn't identifying what needs to be fixed, it's scaling the process of actually fixing it across your entire organization," Gary continued, advancing to the next slide.

Slide 16: Scaling and Making Change Happen

"This is where cloud governance transitions from a technical problem to an organizational transformation challenge."

"As you implement these approaches in your own organizations, you'll face decisions about how to balance control with enablement, how much change teams can absorb, what risk tolerance is appropriate, and what kind of cloud governance culture you want to build," Gary explained. "Through our experience, we've developed principles that guide these decisions."

Cloud Governance Principles

- Start small, scale with success
- Make the right way the easy way
- Enable teams, don't control them
- Build guardrails, not gates
- Fix fast, fail safe
- Automate the routine, humanize the exceptions

Key Implementation Decisions

Gary continued. "As you apply these principles, you'll need to make decisions about..."

- How much change can teams absorb at once?
- What's your risk tolerance for different environments?
- How much freedom should teams have to adapt policies?
- When should you prevent vs. detect and correct?
- How long can issues remain unresolved before escalation?
- How do you balance user experience with control requirements?

"Build your Cloud Champions network to drive adoption across business units, invest in education and communication programs, and allow time for proactive solution architecture to find patterns you can automate."

When Manual Processes Become the Bottleneck:

"You may find that managing this process becomes too manual, with too many steps. And that's exactly right, that's why it's not being done effectively today," Gary continued. He displayed the next slide.

- Visibility alone is not enough; invest in a cloud governance platform that can automate the workflows we've discussed
- Focus your human effort on strategy, relationships, and complex problem-solving rather than administrative overhead

"Remember, the goal isn't to automate everything immediately," Gary emphasized. "It's to build sustainable processes guided by core principles that can evolve with your organization's needs."

"These principles help you make consistent decisions when facing the inevitable edge cases, exceptions, and organizational tensions that come with implementing cloud governance at scale."

Slide 17: Conclusion

"I'd like to leave you with this final thought," Gary said, his voice reflecting the confidence that had grown throughout the presentation. "Managing cloud environments isn't about restricting what teams can do in the cloud. It's about creating the conditions where they can innovate safely and effectively."

Looking out at the audience, Gary made eye contact with members of his team scattered throughout the room. Sarah's confident nod, Fran's subtle smile, and even Barry's attentive gaze reflected how far they'd come as a group.

"What we've discovered is that effective cloud governance requires a shift in mindset, from controlling to enabling, from preventing to guiding, from enforcing to influencing."

"The framework I've shared today isn't just theoretical, it's battle-tested through successes and failures, through resistance and collaboration, through crises and capability-building. It emerged from the collective wisdom of teams across our organization who taught us that effective cloud governance is an ongoing journey, not a fixed destination."

Gary took a breath, delivering his final words with conviction.

"This represents a fundamental shift in how we think about cloud adoption. It's not about controlling the cloud, it's about enabling the organization to use it effectively. It's not about enforcing compliance, it's about creating capabilities. It's not about saying no, it's about helping teams say yes responsibly."

"This is what we've been working toward at Goliath. Not just better controls or tools, but a new approach that balances innovation with risk, speed with security, autonomy with alignment."

"This is Governance for the Cloud Age."

About the Authors

Bob Tordella

Bob is a cloud governance leader with over 20 years of experience in infrastructure, operations, and security. As CRO at Turbot, he helps Global 2000 enterprises bring order to the cloud through automation and policy-driven governance. He is also a core contributor to CloudGovernance.org, a community resource focused on the cultural side of cloud transformation. Bob lives with his wife and two kids in New Hope, Pennsylvania, where he is reminded daily that herding clouds is not so different from herding children.

Nathan Wallace

Nathan is the Founder & CEO of Turbot, where he helps teams bring order to the cloud through governance and security. Over the past two decades in enterprise IT and startups, he's built a career around turning messy infrastructure into elegant automation. At Johnson & Johnson, he led the company's transformation into the cloud and DevOps era, scaling change across one of the world's largest enterprises. An Australian now living in New Jersey, Nathan enjoys travel and the occasional over-engineered project, like the time he built a 15-pound hamburger.

Made in the USA
Las Vegas, NV
28 October 2025

33088511R00195